A Concise Introduction to Image Processing using C++

CHAPMAN & HALL/CRC
Numerical Analysis and Scientific Computing

Aims and scope:
Scientific computing and numerical analysis provide invaluable tools for the sciences and engineering. This series aims to capture new developments and summarize state-of-the-art methods over the whole spectrum of these fields. It will include a broad range of textbooks, monographs and handbooks. Volumes in theory, including discretisation techniques, numerical algorithms, multiscale techniques, parallel and distributed algorithms, as well as applications of these methods in multi-disciplinary fields, are welcome. The inclusion of concrete real-world examples is highly encouraged. This series is meant to appeal to students and researchers in mathematics, engineering and computational science.

Editors

Choi-Hong Lai
School of Computing and
Mathematical Sciences
University of Greenwich

Frédéric Magoulès
Applied Mathematics and
Systems Laboratory
Ecole Centrale Paris

Editorial Advisory Board

Mark Ainsworth
Mathematics Department
Strathclyde University

Peter Jimack
School of Computing
University of Leeds

Todd Arbogast
Institute for Computational
Engineering and Sciences
The University of Texas at Austin

Takashi Kako
Department of Computer Science
The University of Electro-Communications

Craig C. Douglas
Computer Science Department
University of Kentucky

Peter Monk
Department of Mathematical Sciences
University of Delaware

Francois-Xavier Roux
ONERA

Ivan Graham
Department of Mathematical Sciences
University of Bath

Arthur E.P. Veldman
Institute of Mathematics and Computing Science
University of Groningen

Proposals for the series should be submitted to one of the series editors above or directly to:
CRC Press, Taylor & Francis Group
4th, Floor, Albert House
1-4 Singer Street
London EC2A 4BQ
UK

Published Titles

A Concise Introduction to Image Processing using C++
Meiqing Wang and Choi-Hong Lai

Grid Resource Management: Toward Virtual and Services Compliant Grid Computing
Frédéric Magoulès, Thi-Mai-Huong Nguyen, and Lei Yu

Numerical Linear Approximation in C
Nabih N. Abdelmalek and William A. Malek

Parallel Algorithms
Henri Casanova, Arnaud Legrand, and Yves Robert

Parallel Iterative Algorithms: From Sequential to Grid Computing
Jacques M. Bahi, Sylvain Contassot-Vivier, and Raphael Couturier

A Concise Introduction to Image Processing using C++

Meiqing Wang
Choi-Hong Lai

CRC Press is an imprint of the
Taylor & Francis Group an **informa** business

A CHAPMAN & HALL BOOK

Chapman & Hall/CRC
Taylor & Francis Group
6000 Broken Sound Parkway NW, Suite 300
Boca Raton, FL 33487-2742

© 2009 by Taylor & Francis Group, LLC
Chapman & Hall/CRC is an imprint of Taylor & Francis Group, an Informa business

No claim to original U.S. Government works
Printed in the United States of America on acid-free paper
10 9 8 7 6 5 4 3 2 1

International Standard Book Number-13: 978-1-58488-897-0 (Hardcover)

This book contains information obtained from authentic and highly regarded sources. Reasonable efforts have been made to publish reliable data and information, but the author and publisher cannot assume responsibility for the validity of all materials or the consequences of their use. The authors and publishers have attempted to trace the copyright holders of all material reproduced in this publication and apologize to copyright holders if permission to publish in this form has not been obtained. If any copyright material has not been acknowledged please write and let us know so we may rectify in any future reprint.

Except as permitted under U.S. Copyright Law, no part of this book may be reprinted, reproduced, transmitted, or utilized in any form by any electronic, mechanical, or other means, now known or hereafter invented, including photocopying, microfilming, and recording, or in any information storage or retrieval system, without written permission from the publishers.

For permission to photocopy or use material electronically from this work, please access www.copyright.com (http://www.copyright.com/) or contact the Copyright Clearance Center, Inc. (CCC), 222 Rosewood Drive, Danvers, MA 01923, 978-750-8400. CCC is a not-for-profit organization that provides licenses and registration for a variety of users. For organizations that have been granted a photocopy license by the CCC, a separate system of payment has been arranged.

Trademark Notice: Product or corporate names may be trademarks or registered trademarks, and are used only for identification and explanation without intent to infringe.

Visit the Taylor & Francis Web site at
http://www.taylorandfrancis.com

and the CRC Press Web site at
http://www.crcpress.com

Table of Contents

Preface   xv

CHAPTER 1 ▪ Basic Concepts of Images   1

  1.1 ANALOGUE SIGNALS   1
  1.2 DIGITAL SIGNALS   3
      1.2.1 Sampling   4
      1.2.2 Quantisation   5
  1.3 GREY-SCALE IMAGES   6
      1.3.1 Resolution   6
      1.3.2 Grey Levels   6
  1.4 COLOUR IMAGES   7
      1.4.1 The RGB Colour Model   9
      1.4.2 The YIQ Colour Model   10
      1.4.3 The YUV Model   11
      1.4.4 The HSI Model   12
            1.4.4.1 *Conversion from the* RGB *Model to the* HSI *Model*   14
            1.4.4.2 *Conversion from the* HSI *Model to the* RGB *Model*   14
      1.4.5 The CMY Model   16
  1.5 IMAGE STORAGE FORMATS   17
      1.5.1 The BMP Format   17
      1.5.2 The RAW Format   18

	1.5.3	The JPEG format	18
	1.5.4	The GIF Format	19
1.6	VIDEO		19
1.7	EXERCISES		20
1.8	REFERENCES		20
1.9	PARTIAL CODE EXAMPLES		21
	Project 1-1: Convert an 8-bit grey-scale image to a binary image		21
	Project 1-2: Convert a 24-bit colour image to its red channel image		22
	Project 1-3: Convert an 8-bit colour image to a grey-scale image		25

CHAPTER 2 ■ Basic Image Processing Tools 29

2.1	CORRELATION OPERATION AND CONVOLUTION OPERATION		30
	2.1.1	Correlation Operations	30
	2.1.2	Convolution Operations	32
2.2	FOURIER TRANSFORM		37
	2.2.1	Continuous Fourier Transform	37
		2.2.1.1 One-Dimensional Continuous Fourier Transform	37
		2.2.1.2 Two-Dimensional Continuous Fourier Transform	38
	2.2.2	The Discrete Fourier Transform	38
	2.2.3	Properties of the Discrete Fourier Transform	39
	2.2.4	The Fast Fourier Transform	41
2.3	THE DISCRETE COSINE TRANSFORM		42
2.4	THE GABOR TRANSFORM		43
2.5	THE WAVELET TRANSFORM		44
	2.5.1	The Continuous Wavelet Transform	44
	2.5.2	The Discrete Wavelet Transform	45

2.6	\multicolumn{2}{l}{FURTHER READING: ORTHOGONALITY AND COMPLETENESS}	46

- 2.6 **FURTHER READING: ORTHOGONALITY AND COMPLETENESS** — 46
  - 2.6.1 Orthogonality — 47
  - 2.6.2 Completeness — 47
- 2.7 **EXERCISES** — 48
- 2.8 **REFERENCES** — 49
- 2.9 **PARTIAL CODE EXAMPLES** — 49
  - Project 2-1: Fourier Transformation — 49
  - Project 2-2: DCT Transformation — 58
  - Project 2-3: Wavelet Transformation and the inverse wavelet transformation — 60

CHAPTER 3 ■ Preprocessing Techniques for Images — 65

- 3.1 **PIXEL BRIGHTNESS (GREY-LEVEL) TRANSFORMATIONS** — 66
  - 3.1.1 Image Enhancement Based on Histogram — 66
    - 3.1.1.1 Histogram — 66
    - 3.1.1.2 Histogram Equalisation — 66
  - 3.1.2 Contrast Stretching — 69
    - 3.1.2.1 Linear Transform — 70
    - 3.1.2.2 The Limiting Linear Transform — 70
- 3.2 **CONCEPTS AND MODELS OF IMAGE PREPROCESSING** — 71
- 3.3 **IMAGE SMOOTHING** — 73
  - 3.3.1 Spatial-Domain Methods — 73
    - 3.3.1.1 Neighbourhood-Averaging Methods — 73
    - 3.3.1.2 Threshold-Averaging Methods — 75
    - 3.3.1.3 Gaussian Filtering — 75
    - 3.3.1.4 Median Filtering — 76
    - 3.3.1.5 Weighted Median Filtering — 77
  - 3.3.2 Frequency-Domain Methods — 78
    - 3.3.2.1 Ideal Low-Pass Filtering — 78

		3.3.2.2 Trapezoidal Low-Pass Filtering	79
		3.3.2.3 Butterworth Low-Pass Filtering	79
3.4	IMAGE ENHANCEMENT		80
	3.4.1	Gradient	80
	3.4.2	Gradient Image	81
	3.4.3	Gradient Operators	81
		3.4.3.1 Roberts Operator	83
		3.4.3.2 Prewitt Operator	83
		3.4.3.3 Sobel Operator	83
		3.4.3.4 Laplacian Operator	84
	3.4.4	High-Pass Filtering	85
		3.4.4.1 Ideal High-Pass Filtering	85
		3.4.4.2 Trapezoidal High-Pass Filtering	87
		3.4.4.3 Butterworth High-Pass Filtering	87
3.5	IMAGE RESTORATION		87
	3.5.1	Image Degradation Model	87
	3.5.2	Image Restoration Based on the Degradation Model	89
		3.5.2.1 Unconstrained Conditional Restoration	89
		3.5.2.2 Constrained Conditional Restoration	90
	3.5.3	Inverse Filtering	91
	3.5.4	Wiener Filtering	92
	3.5.5	Geometric Rectification	93
		3.5.5.1 Spatial Geometric Transforms	94
		3.5.5.2 Confirmation of Pixel Intensities	96
3.6	PROCESSING METHODS USING PARTIAL DIFFERENTIAL EQUATIONS		97
	3.6.1	Diffusion-Based Models	98
		3.6.1.1 The Heat Conduction Model	98
		3.6.1.2 The Anisotropic Diffusion Model	98
	3.6.2	TV-Based Models	102
	3.6.3	Discrete Formats of PDE Models	103

3.7	FURTHER READING		104
3.8	EXERCISES		105
3.9	REFERENCES		106
3.10	PARTIAL CODE EXAMPLES		108
	Project 3-1: Show the Grey-Scale Histogram of an 8-Bit Grey-Scale Image		108
	Project 3-2: Median Filtering		112
	Project 3-3: Gradient Image Obtained by Using Sobel Operator		118
	Project 3-4: Image Restoration Using the Second- and Fourth-Order Partial Differential Equations		124

CHAPTER 4 ▪ Image Segmentation   135

4.1	THRESHOLDING		136
	4.1.1	Semi-Thresholding and Band-Thresholding	136
		4.1.1.1 Semi-Thresholding	136
		4.1.1.2 Band-Thresholding	137
	4.1.2	Histogram-Based Thresholding	137
		4.1.2.1 The Mode Method	138
		4.1.2.2 Adaptive (Local) Method	138
	4.1.3	Optimal (Iterative) Thresholding	138
4.2	EDGE-BASED SEGMENTATION		140
	4.2.1	Edge Image Thresholding	140
	4.2.2	Edge Relaxation	143
	4.2.3	Border Tracing	148
	4.2.4	The Hough Transform	150
4.3	REGION-BASED SEGMENTATION		152
	4.3.1	The Region-Growing Method	152
	4.3.2	The Region-Merging Method	153
	4.3.3	The Region Split-and-Merge Method	154
4.4	FURTHER READING		154
4.5	EXERCISES		155

4.6	REFERENCES	157
4.7	PARTIAL CODE EXAMPLES	158
	Project 4-1: Optimal Thresholding Segmentation	158
	Project 4-2: The Border-Tracing Method	161

CHAPTER 5 ▪ Mathematical Morphology — 167

5.1	SOME BASIC CONCEPTS OF SET THEORY	168
	5.1.1 Sets and Elements	168
	5.1.2 Relationships between Two Sets	168
	5.1.3 Operations Involving Sets	168
5.2	MORPHOLOGY FOR BINARY IMAGES	169
	5.2.1 Binary Morphological Operation	171
	5.2.1.1 Dilation Operation	171
	5.2.1.2 Erosion Operation	173
	5.2.1.3 Opening and Closing Operations	175
	5.2.1.4 Hit-or-Miss Transformation	175
	5.2.2 Applications of Binary Morphological Operations	176
	5.2.2.1 Thinning and Thickening	176
	5.2.2.2 Skeleton Method	177
5.3	MORPHOLOGY FOR GREY-SCALE IMAGES	178
	5.3.1 Basic Grey-Scale Morphological Operations	178
	5.3.1.1 Dilation Operation	178
	5.3.1.2 Erosion Operation	182
	5.3.2 Applications of Grey-Scale Morphological Operations	183
5.4	FURTHER READING	184
5.5	EXERCISES	185
5.6	REFERENCES	185
5.7	PARTIAL CODE EXAMPLES	186
	Project 5-1: Binary Erosion	186
	Project 5-2: Binary Skeleton	195

Chapter 6 ■ Image Compression			199
6.1	IMAGE FIDELITY METRICS		200
6.2	LOSSLESS COMPRESSION		201
	6.2.1	Huffman Encoding	201
	6.2.2	Runlength Encoding	206
6.3	LOSSY COMPRESSION		207
	6.3.1	Predictive Compression Methods	207
	6.3.2	Vector Quantisation	208
	6.3.3	Wavelet Compression	212
	6.3.4	Fractal Compression	217
6.4	IMAGE COMPRESSION STANDARDS: JPEG AND MPEG		220
	6.4.1	The JPEG Standard	221
		6.4.1.1 DCT-Based Method	221
		6.4.1.2 Predictive Method	221
	6.4.2	The MPEG Standard	222
6.5	FURTHER READING		223
6.6	EXERCISES		224
6.7	REFERENCES		225
6.8	PARTIAL CODE EXAMPLES		226
	Project 6-1: Huffman Encoding		226
	Project 6-2: Fractal Image Compression		236
INDEX			243

Preface

With the development of computer technology, there are many applications benefiting industry and real life gained from studying image processing. These applications include, for example, digital TV, medical images, remote sensing, automatic surveillance, traffic surveillance, industry product detecting, etc. Modern life benefits from understanding some of the basic concepts and fundamental processing tools of images. In general, digital imaging technology can be divided into three categories: image processing, image analysis, and image understanding. The output of image processing is also expressed as images, such as results produced by image smoothing or image enhancement. On the other hand, the output of image analysis and image understanding provides description about images, such as edge detection, image recognition of image analysis, and machine vision of image understanding.

This book covers the fundamental concepts of image processing and some of the related mathematical tools. The main aim of this book is to provide clear concepts and algorithms for image preprocessing, including image smoothing, image enhancement, and image restoration, instead of the mathematical rigor of the subject. Some related topics, including image segmentation and image compression, are also introduced in the book. A touch of mathematical morphology is also included in the book as a new image processing tool. In addition, this book includes state-of-the-art methodologies, such as fractal and wavelet compression algorithms, and an image restoration method based on PDE.

This book may be used as a textbook for a term course suitable for senior undergraduate or junior graduate students. The mathematical concepts introduced in the book are made to an appropriate level as well. All algorithms described in the book are illustrated with code implementation. There are many images in the book used to compare the results of different methods.

In addition, many examples are used to illustrate the mathematical concepts in image processing, which are made easy to understand.

This text also aims to provide a shortcut, do-it-yourself text at a suitable mathematical rigor with plenty of code implementation. Students may modify codes to build their own image analysis tool. The book suits students at the level described above and researchers who need to have a concise and clear view of state-of-the-art image processing methodology, as well as coding examples.

The book has been completed with the help of many colleagues and graduate students. Chen Fei provided partial materials of image restoration methods based on PDE; Cheng Hang collected important materials of the chapter on image compression; Zhuang Zhijun supplied the code implementation. We would also like to extend our appreciation for the help given by Lin Jin and Liu Rong, Huang Chensi, Chen Yanjia, Liu Xiaoyang, and Guo Shumin for their efforts in editing various parts of this book.

CHAPTER 1

Basic Concepts of Images

An image may be considered as a two-dimensional signal function defining the brightness or hue or both at the real coordinates (*x*, *y*). Brightness and hue may be represented by means of a real number or an integer, depending on the formation process of an image from the signal emitted from the object. Several important concepts and tools related to images and signals are briefly introduced in this chapter.

1.1 ANALOGUE SIGNALS

An analogue signal is a continuous variation of certain intensity information with respect to time and can be used to show the time variation of the information. There are simple signals and composite signals that are made up by superimposing simple signals.

A sine wave is a typical example of simple signals and depends on three parameters: amplitude, frequency, and phase angle. The definitions of these three parameters are listed here:

1. *Amplitude:* The amplitude refers to the maximum intensity of a wave. It is denoted as *A*.

2. *Period and frequency:* The period of a wave is the time for it to travel one complete wave cycle. It is denoted as T and measured in units of seconds. The number of cycles per second (cps) is the wave's frequency, which is denoted as f and measured in the interchangeable unit hertz (Hz). Period is the reciprocal of frequency, that is, $T = 1/f$, and vice versa.

3. *Phase:* The phase denotes the position that a wave offsets at the origin of the temporal axis. It is usually denoted as an angle ϕ.

Figure 1.1a shows the intensity and time variation of a single sine pulse represented by $I = A\sin(2\pi ft + \phi)$, where A is the amplitude, f is the frequency, t is time, and ϕ is a certain phase angle. It is also known as the time-domain representation of the signal. The temporal domain given in Figure 1.1a only illustrates the relation between the amplitude and time, but the phase and frequency are not presented in the figure. To show the relationship of amplitude, frequency, and phase, one can use a frequency-domain plot [1]. There are two kinds of frequency-domain plots: amplitude–frequency-domain plots and phase–frequency-domain plots. The former is more frequently used. Figure 1.1b shows the frequency-domain plot with respect to the sine wave.

Any composite analogue signal may be represented as a combination of simple sine/cosine waves with different frequencies, phases, and amplitudes. Figure 1.2 shows a composite analogue signal that is a combination of three simple sine waves.

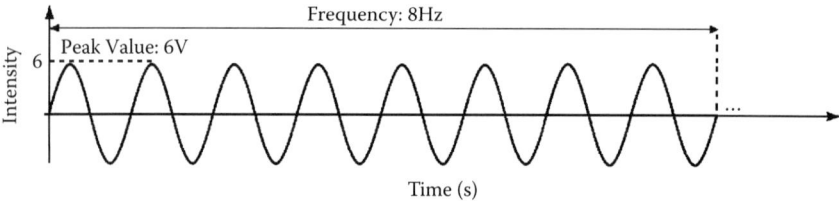

(a) A sine wave in the time-domain (peak value: 6V, frequency: 8Hz).

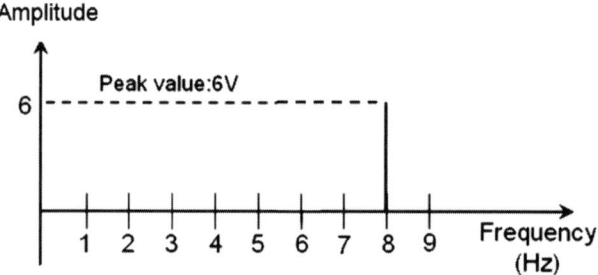

(b) The same sine wave in the frequency domain (peak value: 6V, frequency: 8Hz).

FIGURE 1.1 A simple analogue signal of a sine wave (peak value: 6 V, frequency: 8 Hz) in (a) time domain and (b) frequency domain.

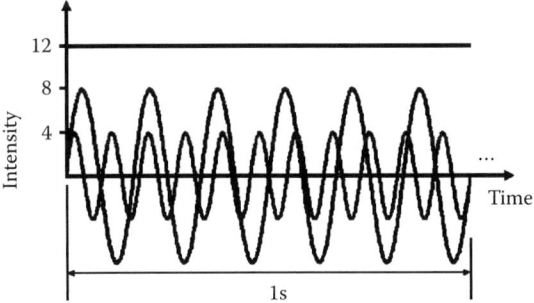

(a) Time-domain representation of three sine waves with frequency 0, 6, and 11.

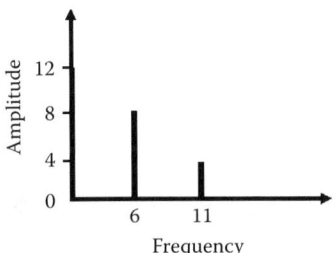

(b) Frequency-domain representation of the same three signals

FIGURE 1.2 The combination of three signals (sine waves with frequency 0, 6, and 11) in (a) time domain and (b) frequency domain.

1.2 DIGITAL SIGNALS

Signals that can be processed by a computer are known as *digital signals*. Analogue signals move back and forth between two peaks in a continuous form. Digital signals maintain a fixed value for a short period of time before changing to another value. The main characteristic of a digital signal is that the intensity is restricted within a limited number of defined values, that is, it is discrete rather than continuous. Figure 1.3 depicts a typical digital signal showing a fixed value within a short period of time.

In order to store and process analogue signals, one can use digital signals to approximate them. For example, to create digital music from analogue music on a cassette tape to play or save on a computer, one needs to convert the analogue signals into digital signals, which involves two processes: sampling and quantisation.

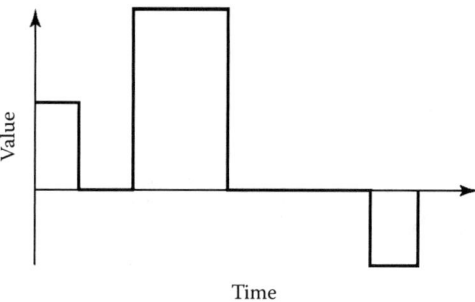

FIGURE 1.3 A digital signal.

1.2.1 Sampling

Sampling is the process of measuring and preserving the signal intensity at a given time. When analogue signals are being converted into digital signals, suitable intervals should be chosen on the discrete space to which the signal function defined in the continuous space is converted. Figure 1.4

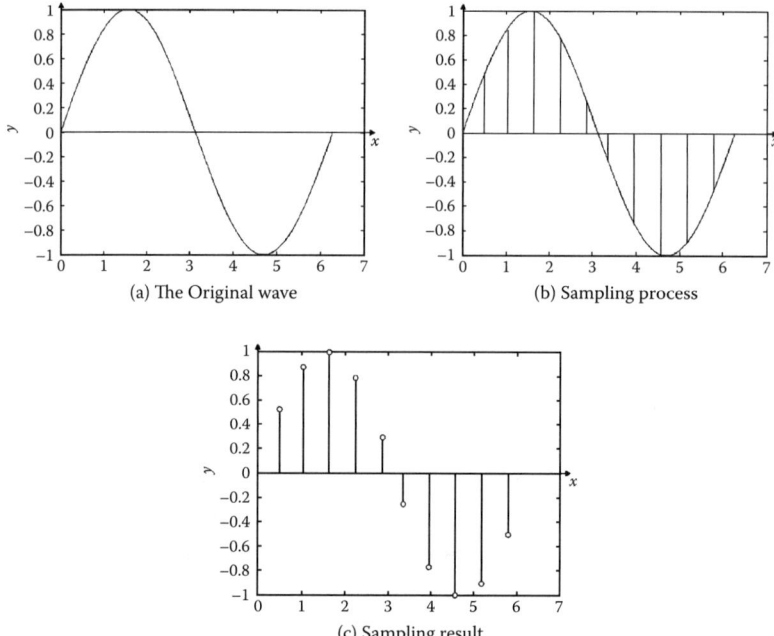

FIGURE 1.4 A typical sampling process: (a) the original wave, (b) sampling process, and (c) sampling result.

illustrates the idea of discrete spatial coordinates and the corresponding signal intensity after a sampling process.

1.2.2 Quantisation

Quantisation replaces a range of values to a single quantum value in order to save storage space. After sampling, the function value at each of the discrete points is a real number. However, only a finite number of quantum values are used to represent the samples. It is possible to use a 2-bit, 4-bit, 8-bit, 16-bit, or 24-bit memory to store these quantum values, depending on the capacity of the chip. For example, a 2-bit memory can store $2^2 = 4$ integers, and an 8-bit memory can store $2^8 = 256$ integers. Given the number of bits, the signal intensity in real number at a particular set of coordinates is mapped to the corresponding quantum value fitted into the available storage space. This process is known as quantisation. There are two types of quantisation, namely, *uniform* and *nonuniform*.

The process of uniform quantization is described as follows. Suppose the amplitude of a signal is A and the corresponding storage is b bits, then $[0, A]$ is divided into 2^b intervals of uniform length. Each interval is called a *level*, and the length of an interval is called the *quantisation step*. There are 2^b quantum values to be stored by using b bits of memory representing the 2^b intervals. The signal intensity at a given coordinate that falls into a particular interval can be approximated by using the corresponding quantum value in the interval. Figure 1.5 shows an example of a uniform quantisation process.

In nonuniform quantisation, the length of one interval, that is, the quantisation step, is not necessarily equal to that of another interval.

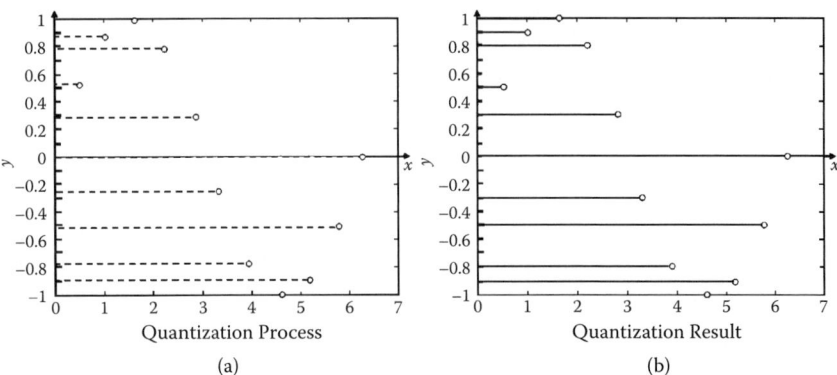

FIGURE 1.5 The uniform quantization process: (a) quantization process, (b) quantization result.

1.3 GREY-SCALE IMAGES

Image formation using sensors and other image acquisition equipment denote the brightness or intensity I of the light of an image as a two-dimensional continuous function $f(x, y)$, where (x, y) denotes the spatial coordinates when only the brightness of light is considered. Sometimes three-dimensional spatial coordinates are used. Images involving only intensity are called *grey-scale images*.

1.3.1 Resolution

Similar to one-dimensional time signals, sampling for images is done in the spatial domain, and quantization is done for the brightness value.

In the sampling process, the domain of an image is divided into N rows and M columns. The region of intersection of a row and a column is known as a *pixel*. The value assigned to each pixel is the average brightness of the region. The position of each pixel is described by a pair of coordinates (x_i, y_j) and may be denoted by means of the indices (i, j), where i and j are integers. For simplicity, $g(x_i, y_j)$ is denoted as $g(i, j)$, where g is a certain property of the region. A grey-scale image, after sampling, is described by an intensity matrix. Throughout this chapter, the simplified notation is used to convey these concepts.

The resolution of a digital signal is the number of pixels presented in the form of number of columns × number of rows. For example, an image with a resolution of 640×480 means that it displays 640 pixels on each of the 480 rows. Some other common resolutions used are 800×600 and 1024×768, among others.

Resolution is one of the most commonly used ways to describe the image quality of a digital camera or other optical equipment. The resolution of a display system or printing equipment is often expressed in number of dots per inch. For example, the resolution of a display system is 72 dots per inch (dpi) or 28 dots per cm.

1.3.2 Grey Levels

Grey levels represent the interval number of quantization in grey-scale image processing. At present, the most commonly used storage method is 8-bit storage. There are 256 grey levels in an 8-bit grey-scale image, and the intensity of each pixel can have a value from 0 to 255, with 0 being black and 255 being white. Another commonly used storage method is 1-bit storage. There are two grey levels, with 0 being black and 1 being white when a binary image, which is frequently used in medical images,

FIGURE 1.6 A grey-scale image.

is being referred to. As binary images are easy to operate, other storage-format images are often converted into binary images when they are used for image enhancement or edge detection. Figures 1.6 and 1.7 show a typical grey-scale image and a binary image, respectively.

1.4 COLOUR IMAGES

The scenery and objects of nature have very rich colour information. Colours are illumination effects caused by light waves having different wavelengths. If a continuous function is used to show a colour image, it may be represented in the form $I = f(x,y,z,\lambda,t)$, where I is the light

FIGURE 1.7 A binary image.

intensity, (x,y,z) are spatial coordinates, λ is the optical wavelength, and t is time. Continuous change in t produces video images, and different wavelengths cause different colours in different pixels.

In general, three characteristics distinguish one colour from another. They are *intensity*, *hue*, and *saturation*. Intensity is used to express the brightness of a colour as discussed previously. Hue is used to describe the colour of a light, identified by its wavelength. For instance, light with a wavelength ranging between 620 and 760 nm is perceived as red, and its wavelength is the largest within the visible light spectrum. On the other hand, light waves with wavelength ranging between 400 and 430 nm are perceived as violet, and its wavelength is the smallest within the visible light spectrum. Figure 1.8 shows the visible spectrum and the colour distribution [2]. Note that the boundaries between different colours in the visible wavelength range are not defined sharply. In essence, each of the seven colours in nature corresponds to a different hue, and each hue corresponds to a different wavelength of light. Saturation is used to describe the strength or freshness of a colour, and it depends on the ratio of white light to colour. The higher the proportion of white light, that is, the lower the proportion of coloured light, the lower the saturation, and vice versa. The value of saturation is expressed as a percentage, and it varies from 0 to 100%. The saturation of pure white light is 0%, and that of a pure colour light is 100%.

It is well known from optical theory [2,3] that each colour with its background in black is considered a combination of red, green, and blue lights. On the other hand, each colour with its background in white can be produced by a certain combination of yellow, cyan, and purple.

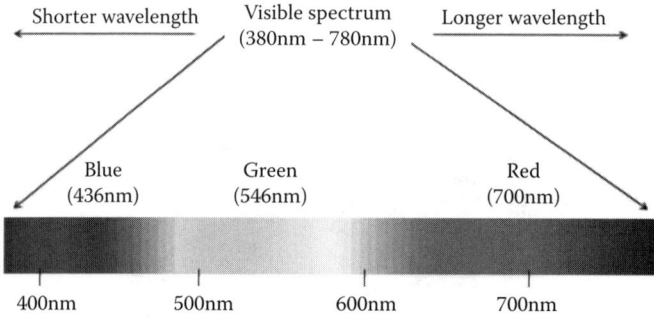

FIGURE 1.8 Visible light spectrum.

Black means there is no colour information. However, pure white light actually contains all colours of the visible spectrum. In the former case of the complete absence of colour, colour is due to additive colour mixing of the three additive primary colours, red, green, and blue. In the latter case of complete colours, colour is due to subtractive colour mixing of the three subtractive primary colours: yellow, cyan, and purple.

1.4.1 The RGB Colour Model

In the RGB colour model, each colour appears in its primary spectral components of red, green, and blue. The colour of a pixel is made up of three components: red, green, and blue (RGB), described by their corresponding intensities. Colour components are also known as *colour channels* or *colour planes*. In the RGB colour model, a colour image can be represented by the intensity function

$$I_{RGB} = (f_R, f_G, f_B) \tag{1.1}$$

where $f_R(x,y)$ is the intensity of the pixel (x, y) in the red channel, $f_G(x,y)$ is the intensity of the pixel (x, y) in the green channel, and $f_B(x,y)$ is the intensity of the pixel (x, y) in the blue channel.

The intensity of each colour channel is usually stored using eight bits, which indicates that the quantization level is 256. That is, a pixel in a colour image requires a total storage of 24 bits. A 24-bit memory can express as $2^{24} = 256 \times 256 \times 256 = 16777216$ distinct colours. The number of colours should adequately meet the display effect of most images. Such images may be called *true colour images*, where information of each pixel is kept by using a 24-bit memory.

Due to computer hardware constraints in the early days, display memory did not meet the requirements of 24-bit storage described here. As a result, the information of each pixel of a colour image could only be stored and displayed by a smaller-size memory, such as an 8-bit or 6-bit memory. Under these circumstances, the palette technology [4] was used. An 8-bit palette technology consists of an RGB colour table with 256 items, each of which is a 24-bit colour information. When the palette technology is used, the storage memory of a pixel is an 8-bit index of palette rather than the 24-bit colour information. The table is stored with pixel indices of the image. If one wishes to use the colour information of a pixel, the index of the pixel is first found from the memory, followed by the colour information corresponding to this index being determined from the palette.

Figure 1.9 shows the images of a 24-bit colour RGB and its three channels.

FIGURE 1.9 The images of a 24-bit colour RGB and its three channels.

(a) A 24-bit colour RGB image
(b) Red channel
(c) Green channel
(d) Blue channel

If only the brightness information is needed, colour images can be transformed to grey-scale images. The transformation [5] can be made by using

$$I_y = 0.30 f_R + 0.59 f_G + 0.11 f_B \qquad (1.2)$$

1.4.2 The YIQ Colour Model

The *YIQ* colour model is often used in colour television (TV) broadcast systems. In this model, a colour image is represented by three components, namely, Y, I, and Q. The Y-channel contains intensity information, whereas the *I* and *Q* channels carry colour information. The advantage of this model is that it removes the correlation between intensity Y and the colour information *I* and *Q*. The human visual system is more sensitive to changes in intensity than to changes in hue or saturation. One can tolerate lower resolution in the components of *I* and *Q* than in Y without perceivable degradation of image quality.

Similar to the *RGB* model, an image defined in the *YIQ* model, I_{YIQ}, can be expressed as

$$I_{YIQ} = (g_Y, g_I, g_Q) \qquad (1.3)$$

by using three functions: $g_Y(x,y)$, the intensity at the pixel (x, y); and $g_I(x,y)$ and $g_Q(x,y)$, the colour information of the pixel (x, y) in the I and Q channels, respectively.

The conversion of an image from the RGB model to the YIQ model is performed through the following matrix vector multiplication [5] of RGB components of each pixel in the RGB model:

$$\begin{bmatrix} g_Y(x,y) \\ g_I(x,y) \\ g_Q(x,y) \end{bmatrix} = \begin{bmatrix} 0.30 & 0.59 & 0.11 \\ 0.60 & -0.27 & -0.32 \\ 0.21 & -0.52 & 0.31 \end{bmatrix} \begin{bmatrix} f_R(x,y) \\ f_G(x,y) \\ f_B(x,y) \end{bmatrix} \quad (1.4)$$

The result contains YIQ components of the same pixel.

1.4.3 The YUV Model

One model commonly used in video encoding and transmission is the *YUV* model. It has one luminance component *Y* and two chrominance components *U* (the difference between the intensity at blue channel and the luminance) and *V* (the difference between the intensity at red channel and the luminance).

The importance of using the *YUV* colour system is that the luminance and the colour information are independent. Images having only *Y* signal components without any *U* and *V* components are grey-scale images varying from black to white. The purpose of using the *YUV* model in colour TV is to take advantage of the luminance signal *Y* in resolving the compatibility problems of colour and a black-and-white TV set. Thus, the black-and-white TV set can also receive colour signals.

Similar to the *YIQ* model, the *YUV* model is also a good representation of images for compression. The reason is that the *YUV* model uses less memory for *U* and *V* component storage and encoding than for the *Y* component. Similarly, an image defined in the *YUV* model I_{YUV} can be expressed as

$$I_{YUV} = (h_Y, h_U, h_V) \quad (1.5)$$

by using the three functions: $h_Y(x,y)$, the intensity at the pixel (x, y); and $h_U(x,y)$ and $h_V(x,y)$, the chrominance information of the pixel (x, y) in the *U* and *V* channels, respectively.

The conversion of an image from the *RGB* model with an 8-bit storage for each colour component to the *YUV* model [5] can be obtained by using the following matrix vector multiplication:

$$\begin{bmatrix} h_Y(x,y) \\ h_U(x,y)-128 \\ h_V(x,y)-128 \end{bmatrix} = \begin{bmatrix} \dfrac{77}{256} & \dfrac{150}{256} & \dfrac{29}{256} \\ \dfrac{-44}{256} & \dfrac{-87}{256} & \dfrac{131}{256} \\ \dfrac{131}{256} & \dfrac{-110}{256} & \dfrac{-21}{256} \end{bmatrix} \begin{bmatrix} f_R(x,y) \\ f_G(x,y) \\ f_B(x,y) \end{bmatrix} \quad (1.6)$$

Note that the second and third components of the right-hand-side vector of Equation 1.6 are often negative values, so the second and third chrominance components of the left-hand side are subtracted by 128 in order to ensure positive numbers, which facilitates encoding.

1.4.4 The HSI Model

As mentioned earlier, colour may be specified by the three quantities hue, saturation, and intensity. The *HSI* model [6] describes the colour of each pixel using the three components: *H*, the hue; *S*, the saturation; and *I*, the intensity or brightness of light. As the *I* component is independent of image colour information, it is possible to avoid the interference of light-and-shade conditions during the analysis of colour. For an image-processing system that requires an estimation of colour characteristics such as colour clustering, etc., one can use the *HSI* model to implement the processing easily.

As discussed above, every colour can be viewed as an additive colour mixing based on the three primary colours (red, green, and blue), and can be described visually using a colour triangle as shown in Figure 1.10a. This colour triangle is an equilateral triangle with three vertices *R*, *G*, and *B*, respectively, representing red, green, and blue. The centre point *W* of the triangle represents white colour. All points along the line *PW* joining any point *P* in the triangle to *W* have the same colour (hue), which is defined by the angle generated by the two vectors *PW* and *RW*. Points along *PW* have different saturations. The nearer to *W* a point is, the lower its saturation.

As the colour triangle is planar, it only reflects the concepts of hue and saturation, but not the concept of intensity. The intensity measurement correlates to the line that goes through the centre of the solid, as shown in Figure 1.10b, and perpendicular to the colour triangle. *H*, *S*, and *I*

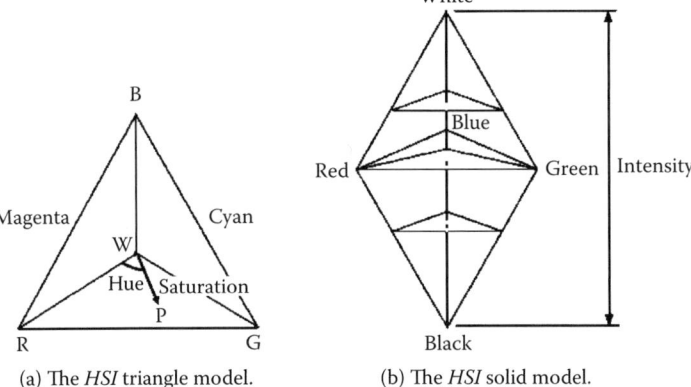

(a) The *HSI* triangle model.   (b) The *HSI* solid model.

FIGURE 1.10 The HSI model depicted by triangle and by solid: (a) the HSI triangle model, and (b) the HSI solid model.

components can be defined using the colour solid. The intensity of a point gradually diminishes to black along the line at the bottom of the solid. On the contrary, the intensity of a point gradually brightens to white along the line to the top of the triangle.

The intensities of the three colour components R, G, and B may be normalised to the range [0,1] as follows:

$$r(x,y) = \frac{f_R(x,y)}{f_R(x,y) + f_G(x,y) + f_B(x,y)} \tag{1.7a}$$

$$g(x,y) = \frac{f_G(x,y)}{f_R(x,y) + f_G(x,y) + f_B(x,y)} \tag{1.7b}$$

$$b(x,y) = \frac{f_B(x,y)}{f_R(x,y) + f_G(x,y) + f_B(x,y)} \tag{1.7c}$$

Notice that $r(x,y), g(x,y), b(x,y) \in [0,1]$, and $r(x,y) + g(x,y) + b(x,y) = 1$. Hence, the preceding transformation actually determines the location of the colour of the pixel at (x,y) in the corresponding colour triangle in Figure 1.10a.

A colour image I_{HSI} in the *HSI* model can be expressed as

$$I_{HSI} = (\varphi_H, \varphi_S, \varphi_I) \tag{1.8}$$

by using three functions: $\varphi_H(x,y)$, the hue of the pixel located at (x, y); $\varphi_S(x,y)$, the saturation of the pixel located at (x, y); and $\varphi_I(x,y)$, the intensity of the pixel located at (x, y).

1.4.4.1 Conversion from the RGB Model to the HSI Model

The formulae used to convert an image from the *RGB* model to the *HSI* model are presented here [7]. The hue function is given by

$$\varphi_H(x,y) = \cos^{-1} \frac{\frac{1}{2}[(f_R(x,y) - f_G(x,y)) + (f_R(x,y) - f_B(x,y))]}{[(f_R(x,y) - f_G(x,y))^2 + (f_R(x,y) - f_B(x,y))(f_G(x,y) - f_B(x,y))]^{1/2}} \quad (1.9)$$

the intensity function is

$$\varphi_I(x,y) = \frac{1}{3}[f_R(x,y) + f_G(x,y) + f_B(x,y)] \quad (1.10)$$

and the saturation function is

$$\varphi_S(x,y) = 1 - 3 \times \frac{\min\{f_R(x,y), f_G(x,y), f_B(x,y)\}}{f_R(x,y) + f_G(x,y) + f_B(x,y)} \quad (1.11)$$

The hue value found by using Equation 1.9 lies in the interval $[0,\pi]$. However, in the colour triangle, the angle of a colour *P* with respect to red colour (the angle between *PW* and *RW*) can be found as an arbitrary value in the interval $[0,2\pi]$. In practice, the size of the angle is determined by the proportion of each colour in its intensity, that is, $b_0 = \frac{b}{\varphi_I}, g_0 = \frac{g}{\varphi_I}$. When $b_0 \le g_0$, the angle lies in the interval $[0,\pi]$, and $\varphi_H(x,y)$ can be calculated by using Equation 1.9. When $b_0 > g_0$, the angle lies in the interval $[\pi, 2\pi]$, leading to $\varphi_H(x,y) = 2\pi - \varphi_H(x,y)$.

Note that when the intensity is zero, that is, $f_R(x,y) + f_G(x,y) + f_B(x,y) = 0$, saturation does not make any sense. On the other hand, when the saturation is zero (the white point), the hue does not make any sense. When the hue is zero, it represents red, and $\frac{2}{3}\pi$ represents green and 2π represents blue.

1.4.4.2 Conversion from the HSI Model to the RGB Model

Conversion of a colour image from the *HSI* model to the *RGB* model is carried out in the colour triangle first. The coordinates of a point in the colour triangle can be expressed in *HSI* values or in normalised *R*, *G*, and *B* values denoted as *r*, *g*, and *b*, respectively. This conversion is achieved simply by computing the *r*, *g*, and *b* values of the colour of the pixel located at (*x*, *y*) from its *HSI* values. These formulae of conversion are related to the position of the colour of the pixel (*x*, *y*) in the colour triangle.

- When $0 < \varphi_H(x,y) < \frac{2}{3}\pi$, which means the colour of the pixel (*x*, *y*) lies in the area enclosed by the red vertex *R*, the green vertex *G*, and the white centre *W* of the colour triangle, the formulae of conversion are as follows:

$$b = \frac{1}{3}[1 - \varphi_S(x,y)]$$

$$r = \frac{1}{3}\left[1 + \frac{\varphi_S(x,y)\cos\varphi_H(x,y)}{\cos\left(\frac{1}{3}\pi - \varphi_H(x,y)\right)}\right] \quad (1.12)$$

$$g = 1 - (b + r)$$

- When $\frac{2}{3}\pi < \varphi_H(x,y) < \frac{4}{3}\pi$, which means the colour of the pixel (*x*, *y*) lies in the area enclosed by the green vertex *G*, the blue vertex *B*, and the white centre *W* of the colour triangle, the formulae of conversion are as follows:

$$\varphi_H = \varphi_H - \frac{2}{3}\pi$$

$$r = \frac{1}{3}[1 - \varphi_S(x,y)]$$

$$g = \frac{1}{3}\left[1 + \frac{\varphi_S(x,y)\cos\varphi_H(x,y)}{\cos\left(\frac{1}{3}\pi - \varphi_H(x,y)\right)}\right] \quad (1.13)$$

$$b = 1 - (r + g)$$

- When $\frac{4}{3}\pi < \varphi_H < 2\pi$, which means the colour of the pixel (*x*, *y*) lies in the area enclosed by the blue vertex *B*, the red vertex *R*, and the white centre W of the colour triangle, the formulae of conversion are as follows:

$$\varphi_H = \varphi_H - \frac{4}{3}\pi$$

$$g = \frac{1}{3}[1 - \varphi_S(x,y)]$$

$$b = \frac{1}{3}\left[1 + \frac{\varphi_S(x,y)\cos\varphi_H(x,y)}{\cos\left(\frac{1}{3}\pi - \varphi_H(x,y)\right)}\right] \quad (1.14)$$

$$r = 1 - (r + g)$$

The conversion from the r, g, and b values at the pixel (x, y) to the actual intensities of the pixel (x, y) of the R, G, and B channels is done as follows:

$$f_R(x,y) = 3r\varphi_I(x,y)$$
$$f_G(x,y) = 3g\varphi_I(x,y) \quad (1.15)$$
$$f_B(x,y) = 3b\varphi_I(x,y)$$

1.4.5 The CMY Model

During the printing of output from a printing device, coloured pigments are deposited on paper, and require employing the subtractive mix-colours theory using the three alternative primary colours: cyan, magenta, and yellow. The *CMY* space is complementary to the *RGB* space because red subtracted from white gives cyan, green subtracted from white gives magenta, and blue subtracted from white gives yellow. Colour images in the *CMY* model may be described as follows [5]:

$$I_{CMY} = (l_C, l_M, l_Y) \quad (1.16)$$

where $l_C(x,y)$, $l_M(x,y)$, and $l_Y(x,y)$ are defined as follows:

$$l_C(x,y) = 1 - \frac{f_R(x,y)}{f_R(x,y) + f_G(x,y) + f_B(x,y)} \quad (1.17a)$$

$$l_M(x,y) = 1 - \frac{f_G(x,y)}{f_R(x,y) + f_G(x,y) + f_B(x,y)} \quad (1.17b)$$

$$l_Y(x,y) = 1 - \frac{f_B(x,y)}{f_R(x,y) + f_G(x,y) + f_B(x,y)} \quad (1.17c)$$

This colour model is used in generating the hardcopy output of colour images, and hence, the inverse conversion from *CMY* to *RGB* is of little practical interest.

1.5 IMAGE STORAGE FORMATS

Digital images are generally stored using the bitmap format. Bitmap, also known as a *bit-mapped image*, describes the colour or intensity of pixels of an image one by one and stores the information in a computer using binary bits. It is different from vector graphics, which is described by using points, lines, and planes in graph processing. Bitmap is appropriate in representing many features of image details, and it can reflect effectively the changes of brightness and darkness, complicated scenes, and colour. Its aim is to show vivid images. Unfortunately, bitmap files are usually large. Another disadvantage of bitmap storage is that fidelity may be reduced and sawtooth may appear when zooming images in or out.

On the other hand, a vector graph consists of some graphic elements such as points, lines, rectangles, polygons, circles, arcs, etc. These elements are obtained by using certain geometrical formulae. As a result, vector drawings are usually of small files. Another advantage of vector graphics is that images will not be distorted during zooming in or out, or during rotation. Their disadvantage is that it is difficult to show the living image effect of rich colour levels. Note that showing vector graphics costs time. Images made up of shapes such as line drawings and illustrations, and free zoom logos and words, are often well suited for vector formats.

Some commonly used image storage formats [8] are discussed in the following sections.

1.5.1 The BMP Format

BMP is the abbreviation of bitmap, and the file storing an image in bitmap format has the suffix .bmp. The *BMP* file is a bit-mapped image format developed by Microsoft® and is the standard image format set by Microsoft for Windows. All image-processing software packages running in the Windows operating system normally support this format.

A BMP file consists of three parts: a bitmap-file header, bitmap information, and a bitmap array. The bitmap-file header explains the storage format and the size of the bitmap. Information such as the width and height of the image, the tag indicating whether or not the image data is compressed, etc., is kept in the bitmap-information part. The bitmap array records the colour values in the RGB model at each pixel of the image. Moreover, if the image is not of true colour, then palette is to be used.

1.5.2 The RAW Format

A file with the suffix .raw is usually used to keep records of electronic level produced when image sensors (charge-coupled device [CCD] or complementary metal-oxide semiconductor [CMOS]) transform light signals into electric signals. The image data stored in the RAW file is just the digitised electric signals captured by a camera image sensor such as CCD. A typical RAW file contains uncompressed or unprocessed pixel data.

RAW format files save the information regarding the best-quality images captured by a CCD that is rich enough for processing later. Different manufacturers produce different permutations and conversion methods for CCD/CMOS and RAW records. Before being processed by common image-processing software, the image in the RAW format needs to be converted into the common image format by using conversion software provided by manufacturers.

RAW format files only record the information of each pixel of an image without a header containing information such as the size of the image. It is easy for the researcher to read the file into an array or some other data structure for processing and then store the data structure to a RAW format file. Therefore, many researchers like to process images in the RAW format. One can use the software tool Photoshop or other image-processing tools to convert images in RAW format to other common formats.

1.5.3 The JPEG format

Another popular format used in image storage and display is the JPEG format, in which files have the suffix .jpg. JPEG is the abbreviation for Joint Photographic Experts Group. JPEG image files use the JPEG standard for image encoding. This compression algorithm is different from that of BMP files. The BMP format uses run-length encoding, which leads to a lossless compression algorithm. However, JPEG is a lossy compression algorithm that will lose some information after decoding. JPEG encoding uses the discrete cosine transform (DCT) technology. These will be introduced in the following chapters. Here, lossless and lossy are related to compression algorithms, and their technical details are further explained in Chapter 6.

1.5.4 The GIF Format

GIF is the abbreviation for graphics interchange format. A GIF format file has the suffix .gif. The format includes some key features that make it a common and valuable format for the Internet. Such features include the high compression ratio and storage of multiple images within a single file allowing a primitive form of animation. However, the maximum storage capacity of each pixel is 8-bit, that is, only a maximum of 256 colours can be referenced within a single GIF image. Hence, GIF format should commonly be used for graphics and images with a few colours such as buttons or black-and-white photos.

1.6 VIDEO

A sequence of continuously varying pictures is known as a *video*. Each picture in the sequence is known as a *frame*. In order for human eyes to see the pictures moving continuously without feeling them to be intermittent, 25 or more frames per second must be displayed.

A video signal is usually created by a video source (e.g., vidicon, VCR, or TV Tuner). To transmit an image, a vertical-synchronous (VSYNC) signal must be generated from the video source first. This signal can be used to reset receiver equipment (e.g., a television set), and guarantees that the display of the new image starts from the top of the screen. After generating the VSYNC signal, the first line of the image from the video source is scanned. When these two steps are completed, a level-synchronisation signal is generated from the video source, and the receiver is reset in order to display the next line from the left of the screen. For each line of the image, a scanning beam and a level-synchronisation pulse signal are emitted from the video source.

Different standards or formats have been established for TV signal transmission and broadcast using different technical parameters. Currently, there are three different formats [9], including NTSC, PAL, and SECAM formats. NTSC (National Television Standard Committee) uses a 525-line standard with 30 frames per second and a pixel aspect ratio of 4:3 as the technical parameters. The technical parameters of PAL (Phase Alternate Line) and SECAM (SEquential Couleur Avec Memoire) standards are 25 frames per second, 625 lines in each frame, and a pixel aspect ratio of 4:3.

Earlier TV receivers could not display at the speed of 25 or 30 frames per second, and flicker could be noticed. In order to resolve this problem, these three standards all employ the interlaced scanning (display) technology. In other words, a screen is partitioned into two fields: the first field contains odd lines of the image, and the second field contains even lines of the image. Odd lines are first scanned and displayed, and then even lines are scanned and displayed. This method improved the stability of image display and reduced flicker. Nowadays, equipment is available that is able to achieve progressive scan and that do not require the interlaced display technology.

1.7 EXERCISES

Q.1 Find out the resolution of your computer monitor, digital camera, or laptop screen.

Q.2 Calculate the number of pixels of an image having a resolution of 1024×768.

Q.3 How many grey levels are there in a grey-scale image stored using a 16-bit memory?

Q.4 Using an image-processing tool, such as Photoshop, convert an image with .bmp format to .raw format. Write a program using C++ to implement this function.

Q.5 A true colour image has the resolution of 800×600. Calculate the sizes of the image files when the image is stored using .bmp format and .raw format.

Q.6 The number of photographs a digital camera can store depends on the storage capacity of the camera, fidelity, and resolution of each photograph. Find the relations among them for your camera.

1.8 REFERENCES

1. B. Forouzan, C. Coombs, and S. C. Fegan, *Introduction to Data Communications and Networking*, McGraw-Hill College, New York, 1997.
2. R. C. Gonzales and R. E. Woods, *Digital Image Processing*, Addison-Wesley, Reading, MA, 1992.
3. M. I. Sobel, *Light*, University of Chicago Press, Chicago, IL, 1989.
4. J. Niederst, *Web Design in a Nutshell*, 2nd edition, O'Reilly & Associates, Inc., Sebastopol, CA, 2001.
5. R. Qiuqi, *Digital Image Processing Science*, Publishing House of Electronic Industry, Beijing, 2001 (in Chinese).

6. H. E. Burdick, *Digital Imaging: Theory and Application*, McGraw-Hill, New York, 1997.
7. M. Sonka, V. Hlavac, and R. Boyle, *Image Processing: Analysis and Machine Vision*, 2nd edition, Thomson Learning and PPTPH, Monterey, CA, 1998.
8. J. Miano, *Compressed Image File Formats: JPEG, PNG, GIF, XBM, BMP*, Addison-Wesley Professional, Boston, MA, 1999.
9. S. J. Solari, *Digital Video and Audio Compression*, McGraw-Hill Professional Publishing, New York, 1997.

1.9 PARTIAL CODE EXAMPLES

Project 1-1: Convert an 8-bit grey-scale image to a binary image

(These codes can be found in CD: Project1-1\ source code\ project1-1View .cpp)

```
#include "stdafx.h"
#include "project1_1.h"
#include "project1_1Doc.h"
#include "project1_1View.h"
void CProject1_1View::OnBinarization()
{
int i,j;
      unsigned char *lpSrc;
      CProject1_1Doc* pDoc = GetDocument();
      ASSERT_VALID(pDoc);
      if (pDoc->m_hDIB == NULL)
            return ;
      LPSTR lpDIB = (LPSTR) ::GlobalLock((HGLOBAL) pDoc->m_hDIB);
      LPSTR lpDIBBits=::FindDIBBits (lpDIB);
      int cxDIB = (int) ::DIBWidth(lpDIB);
// Size of DIB - x
      int cyDIB = (int) ::DIBHeight(lpDIB);
// Size of DIB - y
long lLineBytes = WIDTHBYTES(cxDIB * 8);          //
count the number of
// bytes of the image per line
      for (i = 0; i < cyDIB; i++)
      {
            // per column
            for (j = 0; j < cxDIB; j++)
            {
                  // the pointer pointing to the i-th line and j-th picture element
```

```
                lpSrc = (unsigned char*)lpDIBBits +
lLineBytes * (cyDIB - 1 - i) + j;

                // computing the value of gradation
                if(*lpSrc<122) *lpSrc=BYTE(0);
                else *lpSrc = BYTE(255);
            }
        }
    ::GlobalUnlock((HGLOBAL) pDoc->m_hDIB);
    Invalidate(TRUE);
}
```

Project 1-2: Convert a 24-bit colour image to its red channel image

(These codes can be found in CD: Project1-2 directory\source code\ project1-2View.cpp)

```
include "stdafx.h"
#include "project1_2.h"
#include "project1_2Doc.h"
#include "project1_2View.h"
#ifdef _DEBUG
#define new DEBUG_NEW
#undef THIS_FILE
static char THIS_FILE[] = __FILE__;
#endif
/*****************************************************
*********
* Function name:
* Redchannel()
*
* Parameter:
* HDIB hDIB —the handle of the image
*
* Return Value:
* None
*
* Description:
* Get the red's component of the given image
*
*****************************************************
*******/
void Redchannel(HDIB hDIB)
{
```

```cpp
    LPSTR lpDIB;

    // Get and lock the DIB pointer by the DIB's handle
    lpDIB = (LPSTR) ::GlobalLock((HGLOBAL)hDIB);

    // the pointer pointing to the data area of
DIB's pixel
    LPSTR lpDIBBits;

    // the pointer pointing to the DIB's pixel
    BYTE *lpSrc;

    // the width of image
    LONG    lWidth;
    // the height of image
    LONG    lHeight;

    // the number of byte of image per line
    LONG    lLineBytes;

    // the pointer pointing to the structure body of
BITMAPINFO (Win 3.0)
    LPBITMAPINFO lpbmi;

    // the pointer pointing to the structure body of
BITMAPCOREINFO (Win 3.0)
    LPBITMAPCOREINFO lpbmc;

    // Get the pointer pointing to the structure body
of BITMAPINFO (Win 3.0)
    lpbmi = (LPBITMAPINFO)lpDIB;

    // Get the pointer pointing to the structure body of
// BITMAPCOREINFO (Win 3.0)
    lpbmc = (LPBITMAPCOREINFO)lpDIB;

    // the map table of gradation
    BYTE bMap[256];

    // Compute the map table of gradation
// (save the value of gradation of each colour) and
update the DIB's palette
```

```cpp
        int     i,j;
        for (i = 0; i < 256; i ++)
        {
                // Compute the value of this colour's
gradation
                bMap[i] = (BYTE)(lpbmi->bmiColours[i].
rgbRed);
                // Update the red component of DIB's palette
                lpbmi->bmiColours[i].rgbRed = i;

                // Update the green component of DIB's
palette
                lpbmi->bmiColours[i].rgbGreen = i;

                // Update the blue component of DIB's palette
                lpbmi->bmiColours[i].rgbBlue = i;

                // Update the reserve of DIB's palette
                lpbmi->bmiColours[i].rgbReserved = 0;

        }
        // Find the outset position of the DIB's image pixel
        lpDIBBits = ::FindDIBBits(lpDIB);

        // Get the width of the image
        lWidth = ::DIBWidth(lpDIB);

        // Get the height of the image
        lHeight = ::DIBHeight(lpDIB);

        // count the number of byte of the image per line
        lLineBytes = WIDTHBYTES(lWidth * 8);

        // Replace the colour index of each pixel (change
into the value of gradation
        // according to the map table of gradation)

        // Scan by line
        for(i = 0; i < lHeight; i++)
        {

                // Scan by column
```

```
            for(j = 0; j < lWidth; j++)
            {
                // the pointer pointing to the i-th
line and j-th picture pixel
                lpSrc = (unsigned char*)lpDIBBits +
lLineBytes * (lHeight - 1 - i) + j;

                // Transformation
                *lpSrc = bMap[*lpSrc];
            }
        }

        // Unlocking
        ::GlobalUnlock ((HGLOBAL)hDIB);
}
```

Project 1-3: Convert an 8-bit colour image to a grey-scale image (These codes can be found in CD: Project1-3 directory\source code\project1-3View.cpp)

```
#include "stdafx.h"
#include "project1_3.h"
#include "project1_3Doc.h"
#include "project1_3View.h"
#ifdef _DEBUG
#define new DEBUG_NEW
#undef THIS_FILE
static char THIS_FILE[] = __FILE__;
#endif
/*************************************************
*********
* Function:
* Convert256toGray()
*
* Parameter:
* HDIB hDIB —the picture's handle
*
* Return value:
* None
*
* Description:
```

```
 * Transform the 8 bits colour picture into gradation
picture
 *
 *************************************************************
 *******/
void Convert256toGrey(HDIB hDIB)
{
        LPSTR lpDIB;

        // Get and lock the DIB pointer by the DIB's handle
        lpDIB = (LPSTR) ::GlobalLock((HGLOBAL)hDIB);

        // the pointer pointing to the data area of
DIB's pixel
        LPSTR lpDIBBits;

        // the pointer pointing to the DIB's pixel
        BYTE *lpSrc;
        // the width of image
        LONG   lWidth;
        // the height of image
        LONG   lHeight;

        // the number of byte of image per line
        LONG   lLineBytes;

        // the pointer pointing to the structure body of
BITMAPINFO (Win 3.0)
        LPBITMAPINFO lpbmi;

        // the pointer pointing to the structure body of
BITMAPCOREINFO (Win 3.0)
        LPBITMAPCOREINFO lpbmc;

        // Get the pointer pointing to the structure body
of BITMAPINFO (Win 3.0)
        lpbmi = (LPBITMAPINFO)lpDIB;

        // Get the pointer pointing to the structure body
of BITMAPCOREINFO
// (Win 3.0)
        lpbmc = (LPBITMAPCOREINFO)lpDIB;
```

```
        // the map table of gradation
        BYTE bMap[256];

// Compute the map table of gradation (save the value of
gradation of each colour)
// and update the DIB's palette
        int    i,j;
        for (i = 0; i < 256; i ++)
        {
         // Compute the value of this colour's gradation
             bMap[i] = (BYTE)(0.299 * lpbmi-
>bmiColours[i].rgbRed +

                  0.587 * lpbmi->bmiColours[i].rgbGreen +

                  0.114 * lpbmi->bmiColours[i].rgbBlue
+ 0.5);
             // Update the red component of DIB's palette
             lpbmi->bmiColours[i].rgbRed = i;

             // Update the green component of DIB's
palette
             lpbmi->bmiColours[i].rgbGreen = i;

             // Update the blue component of DIB's palette
             lpbmi->bmiColours[i].rgbBlue = i;

             // Update the reserve of DIB's palette
             lpbmi->bmiColours[i].rgbReserved = 0;

        }
        // Find the outset position of the DIB's image pixel
        lpDIBBits = ::FindDIBBits(lpDIB);

        // Get the width of the image
        lWidth = ::DIBWidth(lpDIB);

        // Get the height of the image
        lHeight = ::DIBHeight(lpDIB);

        // count the number of bit of the image per line
        lLineBytes = WIDTHBYTES(lWidth * 8);
```

```cpp
	// Replace the colour index of each pixel (Transform into the value of gradation
	// according to the map table of gradation)

	// Scan by line
	for(i = 0; i < lHeight; i++)
	{

		// Scan by column
		for(j = 0; j < lWidth; j++)
		{
			// the pointer pointing to the i-th line and j-th picture pixel
			lpSrc = (unsigned char*)lpDIBBits + lLineBytes * (lHeight - 1 - i) + j;

			// Transformation
			*lpSrc = bMap[*lpSrc];
		}
	}

	// Unlocking
	::GlobalUnlock ((HGLOBAL)hDIB);
}
```

CHAPTER 2

Basic Image Processing Tools

As discussed in Chapter 1, a colour image constitutes the three monochromatic components: R, G, and B, each of which may be considered as a grey-scale image as far as processing is concerned. Mathematical tools for grey-scale images can be applied separately to each of the monochromatic components in order to handle colour images, using the same notation as in Chapter 1, where $I = f(x, y)$ denotes the light intensity function of a pixel defined at the coordinates (x, y). Here $f(x, y)$ is a function in the spatial domain. Methods of image processing in spatial domain contain *point operations, local (neighbourhood) operations,* and *global operations.* The result of a point operation is only related to a single pixel. For example, threshold processing for the intensity of a pixel is a point operation. The result of a local operation is related to the neighbouring pixels of a given pixel. In another example, a median filtering has the outcome of a pixel, depending on the intensities of its surrounding neighbouring pixels. A global operation is related to the entire image such as the discrete Fourier transform. A usual neighbourhood includes four or eight neighbouring pixels, as shown in Figure 2.1.

Similar to one-dimensional signals where several properties of images may be easily displayed in the frequency domain, a two-dimensional signal $f(x, y)$ can be broken down into a number of simple signals and expressed as a relation between frequency and amplitude. This decomposition requires the use of Fourier transform, and in image processing,

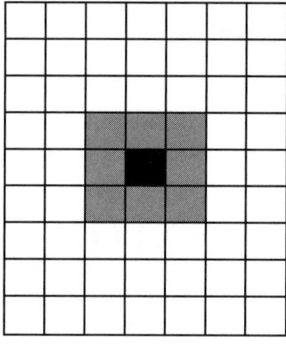

FIGURE 2.1 Two examples of pixel neighbourhoods.

discrete Fourier transform is commonly used. In addition to the discrete Fourier transform, in some cases discrete cosine transform is also used.

This chapter begins with introduction of the concepts of the correlation operation and the convolution operation. The Fourier transform, fast Fourier transform, and the discrete cosine transform are introduced, followed by the Gabor transform and wavelet transform as the basic tools for image processing. In the section on further reading, the concept of orthogonality and completeness of a function set is briefly introduced.

2.1 CORRELATION OPERATION AND CONVOLUTION OPERATION

Correlation and convolution operations are common image-processing tools. These operations are described now.

2.1.1 Correlation Operations

A correlation operation [1] reflects the synchronism or comparability of two signals.

Let $f(t)$ and $g(t)$ be one-dimensional functions in continuous time domain; the correlation R_{fg} between f and g is given by

$$R_{fg}(t) = f(t) \circ g(t) = \int_{-\infty}^{+\infty} f(\alpha) g(t+\alpha) d\alpha \qquad (2.1)$$

Its discrete equivalent operation may be described as follows. Suppose $a(m)$ and $b(m)$ are the corresponding one-dimensional discrete signal sequences, where m is an integer. The correlation between a and b is given by

$$r_{ab}(m) = a \circ b(m) = \sum_{h=-\infty}^{+\infty} a(h)b(h+m) \qquad (2.2)$$

where h is an integer.

Similarly, the correlation between the two-dimensional functions f(x, y) and g(x, y) is given by

$$R_{fg}(x,y) = f(x,y) \circ g(x,y) = \int_{-\infty}^{+\infty}\int_{-\infty}^{+\infty} f(\alpha,\beta)g(x+\alpha,y+\beta)\,d\alpha\,d\beta \qquad (2.3)$$

Suppose the corresponding discretised form for the two-dimensional signals are denoted as a(m, n) and b(m, n), where m and n are integers. The correlation between a and b is given by

$$r_{ab}(m,n) = a(m,n) \circ b(m,n) = \sum_{h=-\infty}^{+\infty}\sum_{l=-\infty}^{+\infty} a(h,l)b(m+h,n+l) \qquad (2.4)$$

where h and l are integers.

Example 2.1 Suppose a(n) and b(n) are two discrete signal sequences in the temporal dimension and are as follows:

$a(0) = 1$, $a(1) = 0.4$, $a(2) = -1$, $a(3) = 0.4$, and all other values of the signal a are zeros.

$b(0) = 0.4$, $b(1) = 1$, $b(2) = 0.4$, $b(3) = -1$, and all other values of the signal b are zeros.

Correlate the two discrete signals.
Solution: Using Equation 2.2, one obtains

$$r_{ab}(m) = a \circ b(m) = \sum_{h=-\infty}^{+\infty} a(h)b(h+m) = \begin{cases} \sum_{h=0}^{3} a(h)b(h+m), & -3 \leq m \leq 3 \\ 0 & \text{otherwise} \end{cases}$$

For $m = -3, -2, -1, 0, 1, 2, 3$, one obtains

$r_{ab}(-3) = a(0)b(-3) + a(1)b(-2) + a(2)b(-1) + a(3)b(0)$
$= a(3)b(0) = 0.4 \times 0.4 = 0.16$

$r_{ab}(-2) = a(0)b(-2) + a(1)b(-1) + a(2)b(0) + a(3)b(1)$
$= a(2)b(0) + a(3)b(1) = -1 \times 0.4 + 0.4 \times 1 = 0$

$r_{ab}(-1) = a(0)b(-1) + a(1)b(0) + a(2)b(1) + a(3)b(2) = a(1)b(0)$
$+ a(2)b(1) + a(3)b(2) = 0.4 \times 0.4 - 1 \times 1 + 0.4 \times 0.4 = -0.68$

$r_{ab}(0) = a(0)b(0) + a(1)b(1) + a(2)b(2) + a(3)b(3)$
$= 1 \times 0.4 + 0.4 \times 1 - 1 \times 0.4 + 0.4 \times (-1) = 0$

$r_{ab}(1) = a(0)b(1) + a(1)b(2) + a(2)b(3) + a(3)b(4) = a(0)b(1) + a(1)b(2) + a(2)b(3)$
$= 1 \times 1 + 0.4 \times 0.4 - 1 \times (-1) = 2.16$

$r_{ab}(2) = a(0)b(2) + a(1)b(3) + a(2)b(4) + a(3)b(5)$
$= a(0)b(2) + a(1)b(3) = 1 \times 0.4 + 0.4 \times (-1) = 0$

$r_{ab}(3) = a(0)b(3) + a(1)b(4) + a(2)b(5) + a(3)b(6) = a(0)b(3) = 1 \times (-1) = -1$

Note that $r_{ab}(1) = 2.16$ is the maximum value obtained in the correlation, and this maximum occurs when $m = 1$. From Figure 2.2 one can see the highest comparability; as the sequence $b(n)$ is shifted left for one unit, the result coincides with the sequence $a(n)$ at most of the points for the case when $m = 1$. ∎

2.1.2 Convolution Operations

Let $f(t)$ and $g(t)$ be one-dimensional functions in continuous time domain; the convolution C_{fg} of the two functions is given by

$$C_{fg}(t) = f(t) * g(t) = \int_{-\infty}^{+\infty} f(\alpha) g(t - \alpha) d\alpha \tag{2.5}$$

Its discrete form may be described by two discrete signal sequences, $a(m)$ and $b(m)$, where m is an integer. The convolution of a and b is given by

$$c_{ab}(m) = a(m) * b(m) = \sum_{h=-\infty}^{+\infty} a(h)b(m-h) \qquad (2.6)$$

where h is an integer.

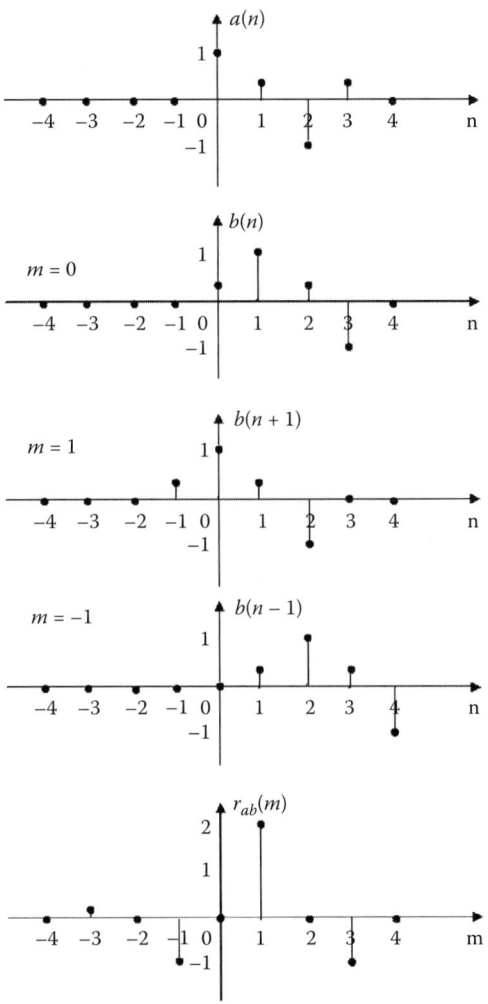

FIGURE 2.2 Results of a correlation operation.

Similarly, the definition of convolution operations between two-dimensional continuous functions $f(x, y)$ and $g(x, y)$ with their corresponding discrete equivalents $a(m, n)$ and $b(m, n)$ is given by

$$C_{fg} = f(x,y)*g(x,y) = \int_{-\infty}^{+\infty}\int_{-\infty}^{+\infty} f(\alpha,\beta)g(x-\alpha, y-\beta)d\alpha\, d\beta$$

$$= \int_{-\infty}^{+\infty}\int_{-\infty}^{+\infty} f(x-\alpha, y-\beta)g(\alpha,\beta)d\alpha\, d\beta \qquad (2.7)$$

$$C_{ab}(m,n) = a(m,n)*b(m,n) = \sum_{h=-\infty}^{+\infty}\sum_{l=-\infty}^{+\infty} a(h,l)b(m-h, y-l)$$

$$= \sum_{h=-\infty}^{+\infty}\sum_{l=-\infty}^{+\infty} a(m-h, y-l)b(h,l) \qquad (2.8)$$

where h and l are integers.

Example 2.2 Calculate the convolution of the two discrete signals as given in Example 2.1.

$a(0) = 1$, $a(1) = 0.4$, $a(2) = -1$, $a(3) = 0.4$, and all other values of the signal a are zeros.

$b(0) = 0.4$, $b(1) = 1$, $b(2) = 0.4$, $b(3) = -1$, and all other values of the signal b are zeros.

Solution: Using Equation 2.6, one obtains

$$C_{ab}(m) = a(m)*b(m) = \sum_{h=-\infty}^{+\infty} a(h)b(m-h) = \begin{cases} \sum_{h=0}^{3} a(h)b(m-h), & 0 \leq m \leq 6 \\ 0 & \text{otherwise} \end{cases}$$

For $m = 0, 1, 2, 3, 4, 5, 6$, one obtains

$$C_{ab}(0) = a(0)b(0) + a(1)b(-1) + a(2)b(-2) + a(3)b(-3) = a(0)b(0) = 1 \times 0.4 = 0.4$$

$$C_{ab}(1) = a(0)b(1) + a(1)b(0) + a(2)b(-1) + a(3)b(-2)$$

$$= a(0)b(1) + a(1)b(0) = 1 \times 1 + 0.4 \times 0.4 = 1.16$$

$$C_{ab}(2) = a(0)b(2) + a(1)b(1) + a(2)b(0) = 1 \times 0.4 + 0.4 \times 1 - 1 \times 0.4 = 0.4$$

$$c_{ab}(3) = a(0)b(3) + a(1)b(2) + a(2)b(1) + a(3)b(0)$$
$$= 1 \times (-1) + 0.4 \times 0.4 - 1 \times 1 + 0.4 \times 0.4 = -1.68$$
$$c_{ab}(4) = a(1)b(3) + a(2)b(2) + a(3)b(1) = 0.4 \times (-1) - 1 \times 0.4 + 0.4 \times 1 = -0.4$$
$$c_{ab}(5) = a(2)b(3) + a(3)b(2) = -1 \times (-1) + 0.4 \times 0.4 = 1.16$$
$$c_{ab}(6) = a(3)b(3) = 0.4 \times (-1) = -0.4$$

Figure 2.3 depicts the corresponding relation of the terms of the two sequences used for computing $c_{ab}(3)$. ■

In an image formation system, the process of converting a physical signal $a(m, n)$ into an electrical signal $c(m, n)$ is usually expressed as a convolution of the input signal and the pulse response of the sensor system. The system may include optical and electronic systems. If each system is linear and shift-invariant (LSI), a convolution model is appropriate. The concepts of linear and LSI systems will be given in the follow-up chapter.

For image processing, the convolution operation is a local operation. The basic idea is to use a window with a given size and shape, known as the supporting window, to scan the entire image. The result is equivalent to the weighted sum of the intensities of the pixels in the window. The weight of each pixel is defined by assigning a value $h(i, j)$ to the location (i, j) in the window. The window with its weights is called the convolution kernel or convolution mask. The matrix h is called a filter and generally defined as 0 outside the window.

The convolution of a filter $h(m, n)$ and an image $a(m, n)$ generates a new image $c(m, n)$, which can be written as follows in terms of the finite sum:

$$c(m,n) = a(m,n) * h(m,n) = \sum_{j=0}^{r} \sum_{k=0}^{s} h(j,k) a(m-j, n-k) \qquad (2.9)$$

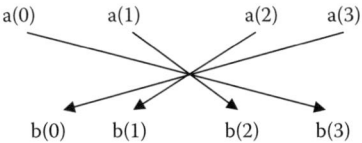

FIGURE 2.3 The corresponding relation used for computing convolution.

In general, the coordinates at the centre of the convolution kernel $h(i, j)$ are set as $(0, 0)$, and hence Equation 2.9 is usually written in the form

$$c(m,n) = a(m,n) * h(m,n) = \sum_{j=-r}^{r} \sum_{k=-s}^{s} h(j,k) a(m-j, n-k) \quad (2.10)$$

If the convolution kernel is a symmetric matrix, that is, $h(-j,-k) = h(j,k)$, then Equation 2.10 becomes

$$c(m,n) = a(m,n) * h(m,n) = \sum_{j=-r}^{r} \sum_{k=-s}^{s} h(j,k) a(m-j, n-k)$$

$$= \sum_{j=-r}^{r} \sum_{k=-s}^{s} h(-j,-k) a(m-j, n-k) = \sum_{j=-r}^{r} \sum_{k=-s}^{s} h(j,k) a(m+j, n+k)$$

$$(2.11)$$

However, there are many asymmetric convolution kernels, in which case Equation 2.11 does not work.

For example, the following discrete convolution kernel is often used to sharpen a given image:

$$h = \begin{bmatrix} h(-1,-1) & h(-1,0) & h(-1,1) \\ h(0,-1) & h(0,0) & h(0,1) \\ h(1,-1) & h(1,0) & h(1,1) \end{bmatrix} = \begin{bmatrix} 0 & -1 & 0 \\ -1 & 5 & -1 \\ 0 & -1 & 0 \end{bmatrix} \quad (2.12)$$

Example 2.3 Suppose there is an 8×8 grey-scale image, the intensity matrix of which is given as follows:

$$a = \begin{bmatrix} 200 & 201 & 202 & 202 & 203 & 202 & 200 & 198 \\ 202 & 203 & 205 & 204 & 204 & 202 & 200 & 197 \\ 205 & 210 & 211 & 212 & 210 & 209 & 208 & 205 \\ 205 & 208 & 210 & 212 & 214 & 210 & 211 & 208 \\ 210 & 212 & 215 & 218 & 217 & 219 & 220 & 218 \\ 212 & 214 & 218 & 220 & 220 & 219 & 218 & 218 \\ 210 & 212 & 213 & 215 & 216 & 216 & 210 & 212 \\ 208 & 208 & 210 & 211 & 212 & 214 & 210 & 210 \end{bmatrix}$$

With a supporting window of size 3 × 3, this example shows the calculation of the intensity of the pixel located at (2, 3), assuming the position of the top left pixel is (1, 1). The convolution of the image $a(m, n)$ and the convolution kernel $h(m, n)$ defined in Equation 2.12 leads to the following result:

$$c(2,3) = \sum_{j=-1}^{1}\sum_{k=-1}^{1} h(j,k)a(2+j,3+k)$$

$$= 0\times 201 + (-1)\times 202 + 0\times 202 + (-1)\times 203 + 5\times 205$$

$$+ (-1)\times 204 + 0\times 210 + (-1)\times 211 + 0\times 212$$

$$= -202 - 203 + 5\times 205 - 204 - 211 = 205$$

in which the intensities of the neighbouring pixels of the pixel located at (2, 3) in the original image are used in the calculation. ∎

2.2 FOURIER TRANSFORM

There are many applications of Fourier transform in image processing, for example, determining the high frequency components of an image function during edge detection and serving as an inverse filter in image restoration.

2.2.1 Continuous Fourier Transform

2.2.1.1 One-Dimensional Continuous Fourier Transform

Suppose $f(t)$ is a function of t, contains only a finite number of discontinuous and extremal points, and is absolute integrable; then the following two integration formulae exist:

$$F(u) = \int_{-\infty}^{+\infty} f(t)e^{-i2\pi ut} dt$$

$$f(t) = \int_{-\infty}^{+\infty} F(u)e^{i2\pi ut} du$$

(2.13)

$F(u)$ is known as the Fourier transform of $f(t)$, and $f(t)$ is known as the inverse Fourier transform of $F(u)$.

Let $w = 2\pi u$, then the Fourier transform in Equation 2.13 can be rewritten as

$$F(w) = \int_{-\infty}^{+\infty} f(t) e^{-iwt} dt$$

$$f(t) = \frac{1}{2\pi} \int_{-\infty}^{+\infty} F(w) e^{iwt} dw$$

(2.14)

2.2.1.2 Two-Dimensional Continuous Fourier Transform

The Fourier transform of a two-dimensional function $f(x, y)$ in the spatial domain is defined as

$$F(u,v) = \int_{-\infty}^{+\infty} \int_{-\infty}^{+\infty} f(x,y) e^{-i2\pi(ux+vy)} dx\, dy$$

$$f(x,y) = \int_{-\infty}^{+\infty} \int_{-\infty}^{+\infty} F(u,v) e^{i2\pi(ux+vy)} du\, dv$$

(2.15)

Similar to one-dimensional cases, $F(u, v)$ is known as the Fourier transform of $f(x, y)$, and $f(x, y)$ is known as the inverse Fourier transform of $F(u, v)$.

2.2.2 The Discrete Fourier Transform

In digital image processing, an image function is often a discretized function leading to a matrix in which each element of the matrix is the intensity of a pixel. Hence, the discrete Fourier transform (DFT) is preferred.

Suppose $a(m), m = 0, 1, 2, \ldots, M-1$, where M denotes the number of discrete points, is a one-dimensional discrete signal. The discrete Fourier transform is defined as

$$A(u) = \sum_{m=0}^{M-1} a(m) e^{-i2\pi\left(\frac{um}{M}\right)}$$

(2.16)

For a two-dimensional discrete signal with $M \times N$ discrete points

$$a(m,n), \quad m = 0, 1, 2, \ldots, M-1; \quad n = 0, 1, 2, \ldots, N-1$$

its discrete Fourier transform is given as follows:

$$A(u,v) = \sum_{m=0}^{M-1}\sum_{n=0}^{N-1} a(m,n) e^{-i2\pi\left(\frac{um}{M}+\frac{vn}{N}\right)}$$

$$= \sum_{m=0}^{M-1}\left(\sum_{n=0}^{N-1} a(m,n) e^{-i2\pi\frac{vn}{N}}\right) e^{-i2\pi\frac{um}{M}}$$

$$= \Gamma_m\{\Gamma_n[a(m,n)]\} \qquad (2.17)$$

where $u = 0,1,2,...,M-1$, $v = 0,1,2,...,N-1$, and Γ_m, Γ_n denote one-dimensional Fourier transforms in the indices of m and n, respectively. Equation 2.17 shows that the two-dimensional Fourier transform can be split into two one-dimensional Fourier transforms.

The inverse discrete Fourier transforms of the preceding two cases are defined as

$$a(m) = \sum_{u=0}^{M-1} A(u) e^{i2\pi\left(\frac{um}{M}\right)}$$

$$a(m,n) = \frac{1}{MN}\sum_{u=0}^{M-1}\sum_{v=0}^{N-1} A(u,v) e^{i2\pi\left(\frac{um}{M}+\frac{vn}{N}\right)} \qquad (2.18)$$

$$m = 0,1,2,...,M-1; n = 0,1,2,...,N-1$$

If the shape of the image is a square, that is, $M = N$, the following symmetry transformation formulae are used:

$$A(u,v) = \frac{1}{N}\sum_{m=0}^{N-1}\sum_{n=0}^{N-1} a(m,n) e^{-i2\pi\left(\frac{um+vn}{N}\right)} \qquad (2.19)$$

$$a(m,n) = \frac{1}{N}\sum_{u=0}^{N-1}\sum_{v=0}^{N-1} A(u,v) e^{i2\pi\left(\frac{um+vn}{N}\right)} \qquad (2.20)$$

2.2.3 Properties of the Discrete Fourier Transform

The discrete Fourier transform has many properties. Some of these properties, which are interesting from an image processing point of view, are listed here. For the sake of simplicity, the operator Γ is used to denote the Fourier transform operation, namely,

$$\Gamma(a(m,n)) = A(u,v)$$

Let $a_1(m,n), a_2(m,n)$ be two discrete image functions, and $A_1(u,v), A_2(u,v)$ be the corresponding Fourier transforms of $a_1(m,n), a_2(m,n)$ according to the definition of Equation 2.19.

1. Linearity:

$$\Gamma\{\alpha a_1(m,n) + \beta a_2(m,n)\} = \alpha \Gamma(a_1(m,n)) + \beta \Gamma(a_2(m,n))$$
$$= \alpha A_1(u,v) + \beta A_2(u,v) \quad (2.21)$$

where α and β are constants.

2. Separability:

$$\Gamma\{a(m,n)\} = \frac{1}{N} \Gamma_m \{\Gamma_n \{a(m,n)\}\} \quad (2.22)$$

3. Shift in the spatial domain:

$$\Gamma\{a(m-\alpha, n-\beta)\} = A(u,v) e^{-i2\pi(\alpha u + \beta v)} \quad (2.23)$$

4. Shift in the frequency domain:

$$\Gamma\{a(m,n) e^{i2\pi(u_0 m + v_0 n)}\} = A(u - u_0, v - v_0)$$

5. The energy conservation theorem (Plancherel theorem, Parseval's theorem):

The discrete Fourier transform according to the definitions of Equations 2.19 and 2.20 satisfies the following energy conservation theorem:

$$\sum_{m=0}^{N-1}\sum_{n=0}^{N-1} |a(m,n)|^2 = \sum_{u=0}^{N-1}\sum_{v=0}^{N-1} |A(u,v)|^2 \quad (2.24)$$

6. The convolution theorem:

Let $H(u, v)$ be the Fourier transform of $h(m, n)$; then the following convolution theorem holds:

$$\Gamma\{h(m,n)*a(m,n)\} = H(u,v)A(u,v) \qquad (2.25)$$

2.2.4 The Fast Fourier Transform

The Fourier transform is a time-consuming computation. For example, the Fourier transform of an original sequence with N points has the computational complexity $O(N^2)$. When N is large, the computing time becomes very high. The Fast Fourier Transform (FFT) [2,3] requires the computational complexity $O(N \log_2 N)$, which significantly reduces the computing time when N is large. For two-dimensional signals such as those in image processing, one-dimensional FFT is required to be applied twice, one in the horizontal direction and the other in the vertical direction. The main idea of the FFT algorithm is to split the original signal sequence with N points into two shorter sequences each with $\frac{1}{2}N$ points that may reduce the number of multiplications in the algorithm. This step may be required to be repeated several times. There are many algorithms for FFT, and each may be achieved by a different butterfly flowchart. For example, the FFT algorithm [4,5] applied to an original sequence with $N = 8$ points using decimation-in-time Radix-2 algorithm can be achieved from the butterfly flowchart shown in Figure 2.4 by taking $W = e^{-\frac{2\pi}{N}i}$.

Example 2.4 Let $x(n)$ be the original sequence of a signal with 8 points, and $X(m)$ the Fourier transform of $x(n)$. Use the butterfly flowchart as shown in Figure 2.4 to compute $X(3)$.
Solution:

$$X(3) = x_2(6) + x_2(7)W^3$$
$$= [x_1(4) + x_1(6)W^6] + [x_1(5) + x_1(7)W^6]W^3$$
$$= \{[x(0) + x(4)W^4] + [x(2) + x(6)W^4]W^6\}$$
$$+ \{[x(1) + x(5)W^4] + [x(3) + x(7)W^4]W^6\}W^3$$
$$= x(0) + x(1)W^3 + x(2)W^6 + x(3)W^1 + x(4)W^4$$
$$+ x(5)W^7 + x(6)W^2 + x(7)W^5$$

The correction can be examined by using DFT as defined in Equation 2.16. ∎

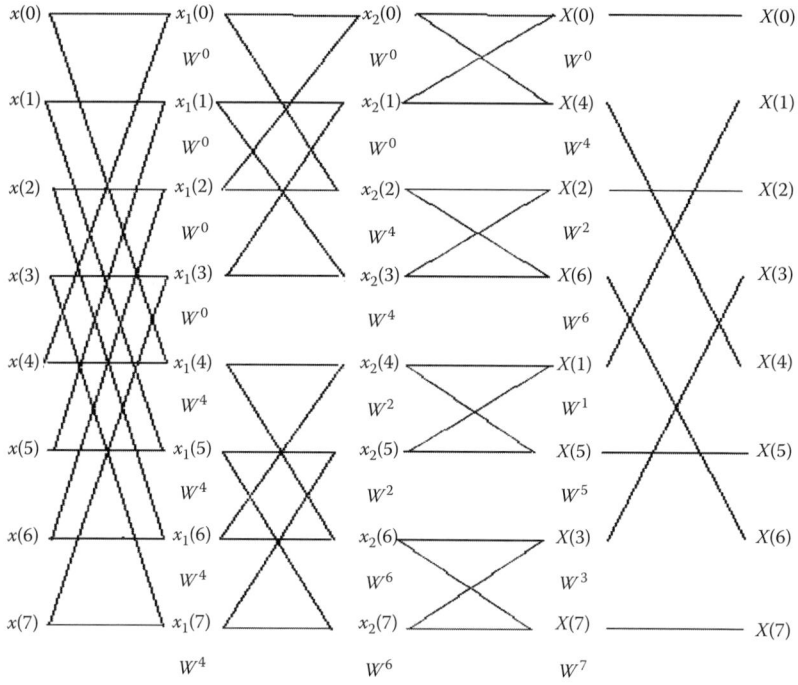

FIGURE 2.4 The eight-point decimation-in-time FFT butterfly flowchart.

2.3 THE DISCRETE COSINE TRANSFORM

The discrete cosine transform (DCT) [6,7] is frequently used in image coding because it involves operations with real numbers in the transform process.

Suppose $f(m, n), m = 0,1,...,N-1;\quad n = 0,1,...,N-1$, is a discrete two-dimensional function. The discrete cosine transform of $f(m, n)$ is

$$F(0,0) = \frac{1}{N}\sum_{m=0}^{N-1}\sum_{n=0}^{N-1} f(m,n)$$

$$F(0,v) = \frac{\sqrt{2}}{N}\sum_{m=0}^{N-1}\sum_{n=0}^{N-1} f(m,n)\cdot\cos\frac{(2n+1)v\pi}{2N}$$

$$F(u,0) = \frac{\sqrt{2}}{N}\sum_{m=0}^{N-1}\sum_{n=0}^{N-1} f(m,n)\cdot\cos\frac{(2m+1)u\pi}{2N}$$

$$F(u,v) = \frac{2}{N}\sum_{m=0}^{N-1}\sum_{n=0}^{N-1} f(m,n)\cdot\cos\frac{(2m+1)u\pi}{2N}\cdot\cos\frac{(2n+1)v\pi}{2N}$$

(2.26)

The formula for the discrete cosine inverse transform is

$$f(m,n) = \frac{1}{N}F(0,0) + \frac{\sqrt{2}}{N}\sum_{v=1}^{N-1} F(0,v)\cos\frac{(2n+1)v\pi}{2N}$$

$$+ \frac{\sqrt{2}}{N}\sum_{u=1}^{N-1} F(u,0)\cos\frac{(2m+1)u\pi}{2N}$$

$$+ \frac{2}{N}\sum_{u=1}^{N-1}\sum_{v=1}^{N-1} F(u,v)\cos\frac{(2m+1)u\pi}{2N}\cos\frac{(2n+1)v\pi}{2N} \quad (2.27)$$

2.4 THE GABOR TRANSFORM

Fourier transform is a global transformation. Any typical value of $F(u)$ in the frequency domain is related to all values of $f(t)$ in the time domain. Similarly, each $f(t)$ in the time domain is the direct sum of each component of $F(u)$ in the frequency domain. This global distribution cannot reflect the local influence. In 1946 Gabor brought forward a windowed Fourier transform [8], which maintains any local influences without losing them. The windowed Fourier transform is now known as the Gabor transform, and it has many similarities with the wavelet transform. The Gabor transform plays an important role in the analysis of nonstationary signals. It is mainly used in character analysis and detection in image processing. The Gabor transform is also known as a short-time Fourier transform.

The Gabor transform of an original signal $f(t)$ related to a given window function $g(t)$ is defined as

$$Gf(w,\tau) = \int_{-\infty}^{+\infty} f(t)g(t-\tau)e^{-iwt}dt \quad (2.28)$$

When the window function is chosen as the Gaussian function, that is,

$$g(t) = g_a(t) = \frac{1}{2\sqrt{\pi a}}e^{-t^2/(4a)} \quad (2.29)$$

it can be shown that the integral of $Gf(w,\tau)$ with τ from $-\infty$ to $+\infty$ is the Fourier transform of $f(t)$:

$$\int_{-\infty}^{+\infty} Gf(w,\tau)d\tau = \int_{-\infty}^{+\infty} f(t)e^{-iwt}dt \quad (2.30)$$

The proof is left as an exercise (see Section 2.7, Q.4) at the end of this chapter for the readers.

One can see that, from the above formula, the instantaneous value $F(w)$ of the signal $f(t)$ in the frequency domain can be decomposed into the superposition of the Gabor transform component $Gf(w,\tau)$. One can now study the influence of local characteristics in the time domain to $F(w)$ by setting the time τ to a specific value.

2.5 THE WAVELET TRANSFORM

The Gabor transform involves a window function that takes into account the influence of a short time interval of the frequency content of the Fourier transform of a given signal. However, in the window function $g_a(t)$ given by Equation 2.29, the size of the window a is a constant for all frequency, which is itself a limitation [9]. When a is small, the higher pitches in the frequency domain are clearer, but the lower pitches are a blur. When a is large, the lower pitches in the frequency domain are better received, but the higher pitches in the time domain from the inverse transform are a blur. Such limitations may be overcome by the use of wavelet transforms.

2.5.1 The Continuous Wavelet Transform

The continuous wavelet transform (CWT) was introduced by Morlet and Grossmann [10] in the early 1980s to overcome the limitation just described. The definition of applying such a transform to the signal $f(t)$ is given by

$$Wf(a,b) = \int_{-\infty}^{+\infty} f(t)\psi_{a,b}(t)\,dt = \int_{-\infty}^{+\infty} f(t)\frac{1}{\sqrt{a}}\psi\left(\frac{t-b}{a}\right)dt \qquad (2.31)$$

where $a > 0$ is the scale parameter, b is the shift parameter, and ψ is a "mother" wavelet. The constant

$$C_\psi = \int_{-\infty}^{+\infty} \frac{|\Psi(\zeta)|^2}{|\zeta|}\,d\zeta$$

is called the admissibility constant, where Ψ is the Fourier transform of ψ. When the admissibility condition $0 < C_\psi < +\infty$ is satisfied, the original signal $f(t)$ can be restored by using the formula

$$f(t) = \frac{1}{C_\psi}\int_0^{+\infty}\int_{-\infty}^{+\infty}\frac{1}{a^2\sqrt{a}}Wf(a,b)\psi\left(\frac{x-b}{a}\right)da\,db \qquad (2.32)$$

Basic Image Processing Tools ■ 45

The admissibility condition implies $\Psi(0)=0$, that is,

$$\int_{-\infty}^{+\infty} \psi(t)dt = 0$$

2.5.2 The Discrete Wavelet Transform

As images are discretised data and are stored as matrices, a discrete wavelet transform [2] is needed in image processing.

In general a and b of Equation 2.31 are discretised as $a = \frac{1}{2^j}$, $b = \frac{k}{2^j}$, where j, k are integers and the equation itself is discretised as

$$Wf\left(\frac{1}{2^j}, \frac{k}{2^j}\right) = \int_{-\infty}^{+\infty} f(t)\psi_{\frac{1}{2^j},\frac{k}{2^j}}(t)dt = \int_{-\infty}^{+\infty} f(t)2^{\frac{j}{2}}\psi(2^j t - k)dt \quad (2.33)$$

The notation in Equation 2.33 is simplified as the following:

$$Wf(j,k) = \int_{-\infty}^{+\infty} f(t)\psi_{j,k}(t)dt = \int_{-\infty}^{+\infty} f(t)2^{\frac{j}{2}}\psi(2^j t - k)dt \quad (2.34)$$

The signal $f(t)$ can be constructed by means of

$$f(t) = \sum_{j=-\infty}^{+\infty}\sum_{k=-\infty}^{+\infty} (Wf(j,k))\psi_{j,k}(t) \quad (2.35)$$

where $\psi_{j,k}(t) = 2^{\frac{j}{2}}\psi(2^j t - k)$ are orthonormal functions and are called wavelet basis functions obtained by shifting and stretching a mother wavelet $\psi(t)$.

For example, the Harr wavelet function defined as follows can be used as a mother wavelet:

$$\psi_H(t) = \begin{cases} 1, & 0 \le t < \frac{1}{2} \\ -1, & \frac{1}{2} \le t < 1 \\ 0, & \text{else} \end{cases}$$

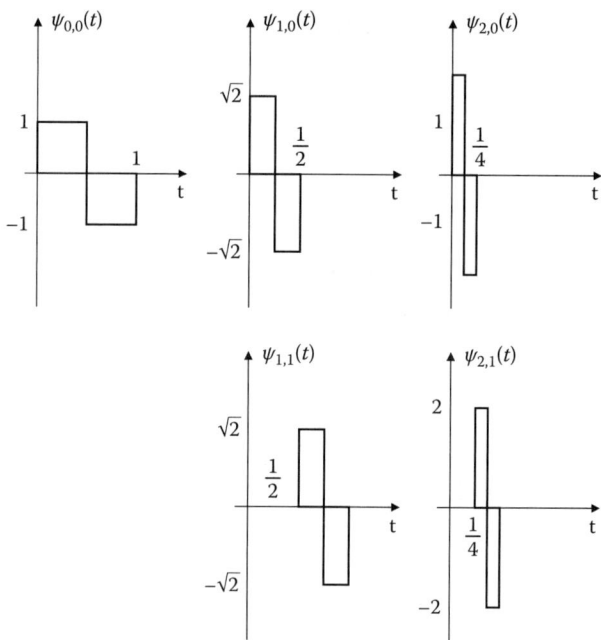

FIGURE 2.5 Some of the Harr wavelet basis functions.

Some wavelet basis functions with their shifted and stretched forms are listed here:

$$\psi_{0,0}(t) = \psi_H(t), \quad \psi_{1,0}(t) = \sqrt{2}\psi_H(2t), \quad \psi_{2,0}(t) = 2\psi_H(4t)$$

$$\psi_{0,1}(t) = \psi_H(t-1), \quad \psi_{1,1}(t) = \sqrt{2}\psi_H(2t-1), \quad \psi_{2,1}(t) = 2\psi_H(4t-1)$$

$$\psi_{0,2}(t) = \psi_H(t-2) \quad \psi_{1,2}(t) = \sqrt{2}\psi_H(2t-2), \quad \psi_{2,2}(t) = 2\psi_H(4t-2)$$

A few of these wavelet basis functions are illustrated in Figure 2.5.

2.6 FURTHER READING: ORTHOGONALITY AND COMPLETENESS

The essence of image transform is to decompose an image function into a weighted sum of a set of basis functions. In order to keep the properties of an image, such as its energy among others, and to ensure that each image function can be decomposed, the function basis must be orthogonal and complete [11,12]. This chapter ends with a revision on the definition of these two important properties.

2.6.1 Orthogonality

Suppose there is a set of real functions $\varphi = \{f_1(t), f_2(t), ..., f_r(t)\}$, which satisfies the following orthogonal property in the interval (t_1, t_2):

$$\int_{t_1}^{t_2} f_i(t) f_j(t) dt = \begin{cases} 0; & i \neq j \\ k; & i = j \end{cases}, \quad i, j = 1, 2, ..., r \quad (2.36)$$

The function set φ is said to be orthogonal in the interval (t_1, t_2). If $k = 1$, it is known as orthonormal.

In the s-dimensional vector space $\Re^s$, the inner product is used to substitute the integral of Equation 2.36. A set of vectors $\psi = \{V_1, V_2, ..., V_r\} \subset \Re^s$ is said to be orthogonal if

$$<V_i, V_j> = \begin{cases} 0; i \neq j \\ k; i = j \end{cases}, \quad i, j = 1, 2, ..., r \quad (2.37)$$

where $V_i, V_j \in \Re^s$.

2.6.2 Completeness

The completeness of the orthogonal function set φ means that, for any real function, $g(t) = 0$ if $g(t)$ is orthogonal to every function in φ. That is,

$$g(t) = 0 \text{ iff } \int_{t_1}^{t_2} f_i(t) g(t) dt = 0 \text{ for each } i = 1, 2, ..., r \text{ and } g(t) \neq f_i(t) \quad (2.38)$$

As for the s-dimensional vector space $\Re^s$, the vector set ψ is complete if for any vector $V \in \Re^s$, $V = 0$ if V is orthogonal to every vector in ψ, that is

$$V = 0 \text{ iff } <V, V_i> = 0 \text{ for each } i = 1, 2, ..., r \text{ and } V \neq V_i \quad (2.39)$$

If the set of vectors ψ is orthonormal and complete, it is easy to show that $r = s$. In this case, ψ is called an orthonormal basis.

The property of completeness ensures that any real function can be written as the weighted sum of the functions in φ, and any s-dimensional vector can be written as the weighted sum of the vectors in ψ.

2.7 EXERCISES

Q.1 The following convolution kernel is often used for smoothing images:

$$h = \begin{bmatrix} \frac{1}{8} & \frac{1}{8} & \frac{1}{8} \\ \frac{1}{8} & 0 & \frac{1}{8} \\ \frac{1}{8} & \frac{1}{8} & \frac{1}{8} \end{bmatrix}$$

Given the 8×8 grey-scale image with the intensity matrix given as below, compute the smoothing result of a by using the convolution kernel h.

$$a = \begin{bmatrix} 200 & 201 & 202 & 202 & 203 & 202 & 200 & 198 \\ 202 & 203 & 205 & 204 & 204 & 202 & 200 & 197 \\ 205 & 210 & 211 & 212 & 210 & 209 & 208 & 205 \\ 205 & 208 & 210 & 212 & 214 & 210 & 211 & 208 \\ 210 & 212 & 215 & 218 & 217 & 219 & 220 & 218 \\ 212 & 214 & 218 & 220 & 220 & 219 & 218 & 218 \\ 210 & 212 & 213 & 215 & 216 & 216 & 210 & 212 \\ 208 & 208 & 210 & 211 & 212 & 214 & 210 & 210 \end{bmatrix}$$

Q. 2 Using Equation 2.26 perform DCT for the image a in Q.1.

Q. 3 Write a program implementing these functions: (i) input of a square image, (ii) output of its Fourier transform coefficients to a file or on screen, and (iii) compare the run times of the standard Fourier transform by using Equation 2.19 and FFT.

Q. 4 Starting from the Gabor transform of a given signal $f(t)$ as shown in Equation 2.28, show that the integration of the l.h.s with τ from $-\infty$ to $+\infty$ is equivalent to the Fourier transform of the same signal.

Q. 5 Show that the Fourier transform and the discrete cosine transform defined in the main text are orthogonal transformations.

2.8 REFERENCES

1. S. W. Smith, *The Scientist and Engineer's Guide to Digital Signal Processing*, California Technical Publishing, 1997.
2. K. R. Castleman, *Digital Image Processing*, 2nd ed., Prentice Hall, Upper Saddle River, NJ, 1996.
3. E. O. Brigham, *The Fast Fourier Transform*, Prentice-Hall, Englewood Cliffs, N J, 1988.
4. H. K. Garg, *Digital Signal Processing Algorithms: Number Theory, Convolution, Fast Fourier Transforms, and Applications*, CRC Press, Boca Raton, FL, 1998.
5. J. S. Walker, *Fast Fourier Transforms*, CRC–Taylor & Francis, 2nd ed., 1996.
6. R. Qiuqi, *Digital Image Processing Science*, Publishing House of Electronic Industry, Beijing, 2001 (in Chinese).
7. W. B. Pennebaker and J. L. Mitchell, *JPEG: Still Image Data Compression Standard*, Springer, New York, 1993.
8. M. J. Bastiaans, Gabor's transform and Zak transform with rational oversampling, in *Signal Processing VIII: Theories and Applications*, Proc. EUSIPCO-96, pp. 2021–2024, ed. G. Ramponi, G. L. Sicuranza, et al., Italy, 1996.
9. W. C. Land and K. Forinash, Time-frequency analysis with the continuous wavelet transform, *American Journal of Physics*, Vol. 66 (9): 794–797, September 1998.
10. A. Grossmann and J. Morlet, Decomposition of Hardy functions into square integrable wavelets of constant shape, *SIAM J. Math. Anal*, Vol. 15, pp. 723–736, 1984.
11. W. Kaplan, *Advanced Calculus*, Addison Wesley, Boston, 4th ed., 1991.
12. G. B. Arfken and H. J. Weber, *Mathematical Methods for Physicists*, Academic Press, 4th ed., 1995.

2.9 PARTIAL CODE EXAMPLES

Project 2-1: Fourier Transformation

(These codes can be found in CD: Project2-1\source code\project2-1 View .cpp)

```
#include "stdafx.h"
#include "project2_1.h"
#include "project2_1Doc.h"
#include "project2_1View.h"
#include "math.h"
/***********************************************************
 *
 * Function name:
 * FFT()
 *
```

```
 * Parameter:
 * complex<double> * TD   - the pointer pointing the
array of time domain
 * complex<double> * FD   - the pointer pointing the
array of frequency range
 * r                       - the power of 2
,which is the times of iteration
 *
 * Return Value:
 * None
 *
 * Description:
 * this function is used to make Fast Fourier Transform
 *
 ************************************************************
 ****************/
VOID WINAPI FFT(complex<double> * TD, complex<double> *
FD, int r)
{
        // the number of transformed dot of Fourier
transform
        LONG    count;

        // Loop variables
        int         i,j,k;

        // Intermediate variable
        int         bfsize,p;

        // the angle
        double angle;

        complex<double> *W,*X1,*X2,*X;

        // compute the number of transformed dot of
Fourier transform
        count = 1 << r;

        // allocate the storage for computing
        W = new complex<double>[count / 2];
        X1 = new complex<double>[count];
        X2 = new complex<double>[count];
```

```
        // calculate the weighting coefficient
        for(i = 0; i < count / 2; i++)
        {
                angle = -i * PI * 2 / count;
                W[i] = complex<double> (cos(angle),
sin(angle));
        }

        // Write the dot of time domain to X1
        memcpy(X1, TD, sizeof(complex<double>) * count);

        // Use the butterfly algorithm for FFT
        for(k = 0; k < r; k++)
        {
                for(j = 0; j < 1 << k; j++)
                {
                        bfsize = 1 << (r-k);
                        for(i = 0; i < bfsize / 2; i++)
                        {
                                p = j * bfsize;
                                X2[i + p] = X1[i + p] + X1[i +
p + bfsize / 2];
                                X2[i + p + bfsize / 2] = (X1[i
+ p] - X1[i + p + bfsize / 2]) * W[i * (1<<k)];
                        }
                }
                X = X1;
                X1 = X2;
                X2 = X;
        }

        // Reordering
        for(j = 0; j < count; j++)
        {
                p = 0;
                for(i = 0; i < r; i++)
                {
                        if (j&(1<<i))
                        {
                                p+=1<<(r-i-1);
                        }
                }
```

```
            FD[j]=X1[p];
        }

        // Freeing
        delete W;
        delete X1;
        delete X2;
}
/***********************************************************
*****************
*
* Function name:
* Fourier()
*
* Parameter:
* LPSTR lpDIBBits - the pointer pointing to the source
of DIB's image
* LONG lWidth - the width of the source image (the
number of pixel)
* LONG lHeight - the height of the source image (the
number of pixel)
*
* Return value:
* BOOL - If succeeded return TRUE else return FALSE
*
* Description:
* Using for making Fourier transform
*
***********************************************************
****************/
BOOL WINAPI Fourier(LPSTR lpDIBBits, LONG lWidth, LONG
lHeight)
{
        // the pointer pointing to the source of DIB's image
        unsigned char*      lpSrc;

        // Intermediate variable
        double dTemp;

        // Cyclic variable
        LONG    i;
        LONG    j;
```

```
        // the width and height of Fourier transformation
(integral power set of 2)
        LONG    w;
        LONG    h;

        int             wp;
        int             hp;

        // the number of byte per line
        LONG    lLineBytes;

        // Compute the number of byte per line
        lLineBytes = WIDTHBYTES(lWidth * 8);

        // Initialization
        w = 1;
        h = 1;
        wp = 0;
        hp = 0;

        // Compute the width and height of Fourier trans-
formation
// (integral power set of 2)
        while(w * 2 <= lWidth)
        {
                w *= 2;
                wp++;
        }

        while(h * 2 <= lHeight)
        {
                h *= 2;
                hp++;
        }

        // Allocation
        complex<double> *TD = new complex<double>[w * h];
        complex<double> *FD = new complex<double>[w * h];

        // Line
        for(i = 0; i < h; i++)
```

```cpp
        {
                // Column
                for(j = 0; j < w; j++)
                {
                        // the pointer pointing to the i-th line and j-th column of DIB's pixel
                        lpSrc = (unsigned char*)lpDIBBits + lLineBytes * (lHeight - 1 - i) + j;

                        // assign the value of time domain
                        TD[j + w * i] = complex<double>(*(lpSrc), 0);
                }
        }

        for(i = 0; i < h; i++)
        {
                // FFT in y's direction
                FFT(&TD[w * i], &FD[w * i], wp);
        }

        // Save the results
        for(i = 0; i < h; i++)
        {
                for(j = 0; j < w; j++)
                {
                        TD[i + h * j] = FD[j + w * i];
                }
        }

        for(i = 0; i < w; i++)
        {
                // FFT in x's direction
                FFT(&TD[i * h], &FD[i * h], hp);
        }

        // Line
        for(i = 0; i < h; i++)
        {
                // Column
                for(j = 0; j < w; j++)
```

```
                    {
                            // Compute the spectra
                            dTemp = sqrt(FD[j * h + i].real() *
FD[j * h + i].real() +
                                    FD[j * h + i].imag() * FD[j *
h + i].imag()) / 100;

                            // Judge whether the dTemp is bigger
than 255
                            if (dTemp > 255)
                            {
                                    // Set 255 to dTemp if it is
bigger than 255
                                    dTemp = 255;
                            }

// the pointer pointing the(i<h/2 ? i+h/2 : i-h/2)-th
line and (j<w/2 ? j+w/2 : j-w/2)-th column of DIB's pixel
                            // Avoid getting the i and j
directly,for moving the origin to the centre
                            //lpSrc = (unsigned char*)lpDIBBits +
lLineBytes * (lHeight - 1 - i) + j;
                            lpSrc = (unsigned char*)lpDIBBits +
lLineBytes *
                                    (lHeight - 1 - (i<h/2 ? i+h/2 :
i-h/2)) + (j<w/2 ? j+w/2 : j-w/2);

                            // Update the source image
                            * (lpSrc) = (BYTE)(dTemp);
                    }
            }

            // Delete the temporary variables
            delete TD;
            delete FD;

            // return
            return TRUE;
}
```

```c
/***********************************************************
*********
* Function name:
* OnFourierTransformation()
*
* Parameter:
* HDIB hDIB - the handle of the image
*
* Return Value:
* None
*
* Description:
* Fourier transform
*
***********************************************************
*******/
void CProject2_1View::OnFourierTransformation()
{
        // Fourier transformation

        // Get the document
        CProject2_1Doc* pDoc = GetDocument();

        // the pointer pointing to DIB's pixel
        LPSTR lpDIB;

        // the pointer pointing to the DIB's pixel
        LPSTR lpDIBBits;

        // Lock DIB
        lpDIB = (LPSTR) ::GlobalLock((HGLOBAL) pDoc->GetHDIB());

   // Find the outset position of the DIB's image pixel
        lpDIBBits = ::FindDIBBits(lpDIB);

        // Judge whether the picture is 8-bpp bits
image(Only deal with the Fourier transformation of 8-bpp
bits image, for the way deal with other types can be
derived from this method)
        if (::DIBNumColours(lpDIB) != 256)
```

```
        {
                // Hint to the user
                MessageBox("It only support Fourier trans-
formation of 8 bits colour picture now!", "Hint from the
system" ,
                        MB_ICONINFORMATION | MB_OK);

                // Unlocking
                ::GlobalUnlock((HGLOBAL) pDoc->GetHDIB());

                // Return
                return;
        }

        // Change the shape of the cursor
        BeginWaitCursor();

        // Invoke the function of Fourier and make FT
        if (::Fourier(lpDIBBits, ::DIBWidth(lpDIB),
::DIBHeight(lpDIB)))
        {
                // Set the flag
                pDoc->SetModifiedFlag(TRUE);

                // Update the views
                pDoc->UpdateAllViews(NULL);
        }
        else
        {
                // Hint for the user
                MessageBox("Allocation is failed!", " Hint
from the system", MB_ICONINFORMATION | MB_OK);
        }

        // Unlocking
        ::GlobalUnlock((HGLOBAL) pDoc->GetHDIB());

        // Reset the shape of the cursor
        EndWaitCursor();
}
```

Project 2-2: DCT Transformation

(These codes can be found in CD: Project2-2\source code\project2-2 View .cpp)

```
#include "stdafx.h"
#include "project2_2.h"
#include "project2_2Doc.h"
#include "project2_2View.h"
#include "math.h"
/*************************************************
 *****************
 *Function Name:
 * DCT()
 *
 * Parameters:
 * double * f                        - the pointer
pointing to time domain
 * double * F                        - the pointer
pointing to frequency range
 * r                                 - the power of 2
 *
 * Return Value:
 * None
 *
 * Description:
 *
 * Use for DCT by the FFT of 2N dots
 *
 *************************************************
 ****************/
VOID WINAPI DCT(double *f, double *F, int r)
{
        // the number of transformed dots of DCT
        LONG    count;

        // Loop variables
        int             i;

        // Intermediate variable
        double dTemp;
```

```
complex<double> *X;

// Compute the number of transformed dots of DCT
count = 1<<r;

// Allocation
X = new complex<double>[count*2];

// Initialisation
memset(X, 0, sizeof(complex<double>) * count * 2);

// Write the dot of time domain to X
for(i=0;i<count;i++)
{
        X[i] = complex<double> (f[i], 0);
}

// Invoke the FFT
FFT(X,X,r+1);

// Adjust coefficient
dTemp = 1/sqrt(count);

// F[0]
F[0] = X[0].real() * dTemp;

dTemp *= sqrt(2);

// F[u]
for(i = 1; i < count; i++)
{
        F[i]=(X[i].real() * cos(i*PI/(count*2)) +
X[i].imag() * sin(i*PI/(count*2))) * dTemp;
}

// Freeing
delete X;
}
```

Project 2-3: Wavelet Transformation and the inverse wavelet transformation

(These codes can be found in CD: Project2-3 directory\source code\project2-3View.cpp)

```
#include "stdafx.h"
#include "project2_3.h"
#include "GlobalApi.h"
#include "project2_3Doc.h"
#include "project2_3View.h"
/*************************************************
*********
* Function name:
* OnWaveletTransform()
*
* Parameter:
* HDIB hDIB - the handle of the image
*
* Return Value:
* None
*
* Description:
* Wavelet Transform
*
*************************************************
*******/
void CProject2_3View::OnWaveletTransform()
{
    // Get the document pointer
    CProject2_3Doc * pDoc = (CProject2_3Doc *)this->GetDocument();
    // change the shape of cursor
    BeginWaitCursor();
    // wavelet transformation
    int rsl = DIBDWTStep(0);
    // reset the shape of cursor
    EndWaitCursor();
    // if the wavelet transformation doesn't work, return directly
    if (!rsl)
        return;
    // set the flag
```

```
            pDoc->SetModifiedFlag(TRUE);
            // update views
            pDoc->UpdateAllViews(FALSE);
}
void CProject2_3View::OnInverseWaveletTransform()
{
            // Get the document pointer
            CProject2_3Doc * pDoc = (CProject2_3Doc *)this->GetDocument();
            // change the shape of cursor
            BeginWaitCursor();
            // wavelet transformation
            int rsl = DIBDWTStep(1);
            // reset the shape of cursor
            EndWaitCursor();
            // if the wavelet transformation doesn't work, return directly
            if (!rsl)
                        return;
            // set the flag
            pDoc->SetModifiedFlag(TRUE);
            // update views
            pDoc->UpdateAllViews(FALSE);
}
BOOL CProject2_3View::DIBDWTStep(int nInv)
{
 // loop variables
            int i,j;
 unsigned char *lpSrc;
            CProject2_3Doc* pDoc = GetDocument();
            LPSTR lpDIB = (LPSTR) ::GlobalLock((HGLOBAL) pDoc->m_hDIB);
            LPSTR lpDIBBits=::FindDIBBits (lpDIB);
            int cxDIB = (int) ::DIBWidth(lpDIB); // Size of DIB - x
            int cyDIB = (int) ::DIBHeight(lpDIB); // Size of DIB - y
            long lLineBytes = WIDTHBYTES(cxDIB * 8); // count the the number of byte of the image per line
            // Get the length and width of image
            int nWidth = cxDIB;
            int nHeight = cyDIB;
```

```
        // Get the biggest number of layers
        int nMaxWLevel = Log2(nWidth);
        int nMaxHLevel = Log2(nHeight);
        int nMaxLevel;
        if (nWidth == 1<<nMaxWLevel && nHeight ==
1<<nMaxHLevel)
                nMaxLevel = min(nMaxWLevel, nMaxHLevel);
        // temporary variables
        double *pDbTemp;
        BYTE   *pBits;

        // if the memory of wavelet transformation wasn't
assigned, allocte it.
        if(!m_pDbImage){
                m_pDbImage = new double[nWidth*nHeight];
                if (!m_pDbImage)   return FALSE;
                // put the image data to m_pDbImage
                for (j=0; j<cyDIB; j++)
                {
                        pDbTemp = m_pDbImage + j*cxDIB;
                        for (i=0; i<cxDIB; i++)
                        {
                                // the pointer pointing to the
i-th line and j-th picture element
                                lpSrc = (unsigned char*)lpDIBBits +
lLineBytes * (cyDIB - 1 - j) + i;
                                pDbTemp[i] = *lpSrc;
                        }
                }
        }

        // wavelet transformation(or inverse wavelet
transformation)
        if (!DWTStep_2D(m_pDbImage, nMaxWLevel-m_nDWTCur-
Depth, nMaxHLevel-m_nDWTCurDepth,
                                        nMaxWLevel, nMax-
HLevel, nInv, 1, m_nSupp))
                return FALSE;
        // if it's inverse transformation ,the number of
layers minus 1
        if (nInv)
                m_nDWTCurDepth --;
```

```
        // else adds 1
        else
            m_nDWTCurDepth ++;

        // copy the data to the former CDib and transform
into the right type
        int lfw = nWidth>>m_nDWTCurDepth, lfh =
nHeight>>m_nDWTCurDepth;
        for (j=0; j<nHeight; j++)
        {
            pDbTemp = m_pDbImage + j*cxDIB;
            pBits = (unsigned char*)lpDIBBits + lLine-
Bytes * (nHeight - 1 - j);
            for (i=0; i<nWidth; i++)
            {
                if (j<lfh && i<lfw)
                    pBits[i] =
FloatToByte(pDbTemp[i]);
                else
                    pBits[i] =
BYTE(FloatToChar(pDbTemp[i])^ 0x80);
            }
        }
        // Return
        return TRUE;
}
```

CHAPTER 3

Preprocessing Techniques for Images

When an image is received or transmitted, a variety of factors will inevitably affect it or interfer with it so that the original specification of the image cannot be retained. As a result, image preprocessing, which includes smoothing, enhancement, and restoration, is needed before one can use the image. The main objective of enhancement is to process an image in such a way that the resulting image becomes more suitable for research purposes and other applications than the original. However, the goal of restoration is to reconstruct or recover an image or part of an image that has been degraded or distorted compared to the original.

This chapter begins with an introduction to pixel brightness and certain transformations related to the analysis of images. Concepts and models of image processing are introduced in Section 3.2. These concepts lead to various image-processing techniques, including image smoothing, enhancement, and restoration, which are introduced in Sections 3.3, 3.4, and 3.5, respectively. Finally, a discussion on the use of partial differential equations in image processing is provided in Section 3.6.

3.1 PIXEL BRIGHTNESS (GREY-LEVEL) TRANSFORMATIONS

3.1.1 Image Enhancement Based on Histogram

A picture becomes dim without enough brightness or exposure to light in the photographing procedure. Brightness can be observed by means of the histogram of an image with a grey level, and the distribution of brightness can be improved by using histogram equalisation.

3.1.1.1 Histogram

A histogram may be used to show the probability of the occurrence of certain grey levels [1,2]. Let r be the normalised grey level, that is, $0 \le r \le 1$, such that $r=0$ and $r=1$ represents the darkest and brightest points, respectively. Let $p_r(r)$ be the probability density function with respect to r. A typical probability density function $p_r(r)$ is depicted in Figure 3.1.

The probability density function should be discretised because the grey levels in an image are discrete. Suppose that the number of pixels and grey levels in an image are N and L, respectively. Let r_k denote the discrete normalised grey levels of an image such that $0 \le r_k \le 1$, $k = 0, 1, ..., L-1$, n_k is the number of pixels in the image that has the grey level r_k, and $\sum_{k=0}^{L-1} n_k = N$. The probability density when $r = r_k$ can be estimated by

$$p_r(r_k) = \frac{n_k}{N} \qquad (3.1)$$

The histogram of the 8-bit image of Lena obtained by using Equation 3.1 is shown in Figure 3.2.

3.1.1.2 Histogram Equalisation

Histogram equalisation ensures that the original probability density function $p_r(r_k)$ has the same value for each of the single grey levels r_k, that is, the image has the same number of pixels at every grey level. Such a method can be used to compare different images from different environments.

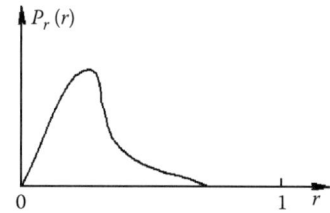

FIGURE 3.1 A probability density sketch map.

 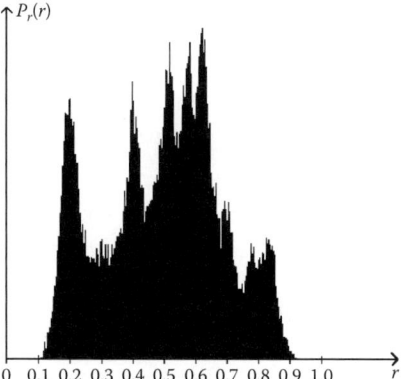

FIGURE 3.2 The source 8-bit image of Lena and its histogram map after normalisation.

Let s be the variable denoting the grey level after equalisation, and its probability density is $p_s(s)=1$. The transformation from the variable r to s is obtained by using $s=T(r)$ and is based on the concept that the same probability distribution holds, that is,

$$\int_0^{T(r)} p_s(s)ds = \int_0^r p_r(r)dr \tag{3.2}$$

After equalisation, that is, $p_s(s)=1$, the integral on the left-hand side of Equation 3.2 is equal to $T(r)$. Hence, the transformation T can be obtained as

$$T(r) = \int_0^r p_r(r)dr \tag{3.3}$$

The discretised formula is given by

$$s_k = T(r_k) = \sum_{j=0}^{k} p_r(r_j) = \sum_{j=0}^{k} \frac{n_j}{N}, k=0,1,...,L-1 \tag{3.4}$$

Example 3.1 A grey-scale image has 16 × 16 pixels. Assuming that the intensity of each pixel in the image requires three bits of storage, the grey levels are 0, 1, 2, 3, 4, 5, 6, and 7. Normalising the grey levels leads to

$$r_0=0, r_1=\frac{1}{7}, r_2=\frac{2}{7}, r_3=\frac{3}{7}, r_4=\frac{4}{7}, r_5=\frac{5}{7}, r_6=\frac{6}{7}, r_7=1$$

The distribution of pixels at each of the preceding grey levels and their corresponding probability densities are given as follows:

$$n_0 = 10, n_1 = 30, n_2 = 50, n_3 = 40, n_4 = 60, n_5 = 30, n_6 = 20, n_7 = 16$$

$$p_r(r_0) = \frac{10}{256}, p_r(r_1) = \frac{30}{256}, p_r(r_2) = \frac{50}{256}, p_r(r_3) = \frac{40}{256}$$

$$p_r(r_4) = \frac{60}{256}, p_r(r_5) = \frac{30}{256}, p_r(r_6) = \frac{20}{256}, p_r(r_7) = \frac{16}{256}$$

Obtain the grey levels after equalisation.

Solution: Using Equation 3.4, r_k can be converted into the new grey level s_k, and $k = 0, 1, \ldots, 7$, listed as follows:

$$s_0 = p_r(r_0) = \frac{10}{256} \approx 0.039$$

$$s_1 = p_r(r_0) + p_r(r_1) = \frac{10}{256} + \frac{30}{256} = \frac{40}{256} \approx 0.156$$

$$s_2 = p_r(r_0) + p_r(r_1) + p_r(r_2) = \frac{10}{256} + \frac{30}{256} + \frac{50}{256} = \frac{90}{256} \approx 0.352$$

$$s_3 = p_r(r_0) + p_r(r_1) + p_r(r_2) + p_r(r_3) = \frac{10}{256} + \frac{30}{256} + \frac{50}{256} + \frac{40}{256} = \frac{130}{256} \approx 0.508$$

Similarly, one obtains

$$s_4 = \frac{190}{256} \approx 0.742, s_5 = \frac{220}{256} \approx 0.859, s_6 = \frac{240}{256} \approx 0.938, s_7 = \frac{256}{256} = 1$$

If s_k is stored by means of a 3-bit storage, these values need to be further approximated as

$$s_0 \approx 0; s_1 \approx \frac{1}{7}; s_2 \approx \frac{2}{7}; s_3 \approx \frac{4}{7}; s_4 \approx \frac{5}{7}; s_5 \approx \frac{6}{7}; s_6 \approx 1; s_7 \approx 1$$

Then the new grey levels are

$$t_0 = 0; t_1 = \frac{1}{7}; t_2 = \frac{2}{7}; t_3 = \frac{4}{7}; t_4 = \frac{5}{7}; t_5 = \frac{6}{7}; t_6 = 1$$

A new image is obtained after converting the original grey levels, according to the following mapping:

$$r_0 \to t_0, r_1 \to t_1, r_2 \to t_2, r_3 \to t_3, r_4 \to t_4, r_5 \to t_5, r_6 \to t_6, r_7 \to t_6$$

The effect of this transformation is summarised here. If the intensity of the pixel (x, y) of the original image is the grey level $\frac{1}{7}$, the intensity of the same pixel of the new image is still the grey level $\frac{1}{7}$. If the intensity of the pixel (x, y) of the original image is the grey level $\frac{3}{7}$, the intensity of the same pixel in the new image is the grey level $\frac{4}{7}$. Figure 3.3 shows the histogram before and after equalisation of the image of Lena depicted in Figure 3.2. ∎

3.1.2 Contrast Stretching

Very often one needs to enhance the contrast of an image. Consider an 8-bit grey-scale image such that the intensity of the pixels can be any grey level of the image and is an integer between 0 and 255. If the maximum intensity value in it is only 150, the image has low contrast and can be observed from the histogram. In order to enhance the contrast of the image, one can directly perform transformations of the intensities of the pixels. Methods often used in such transformations are the linear transform and the limiting linear transform.

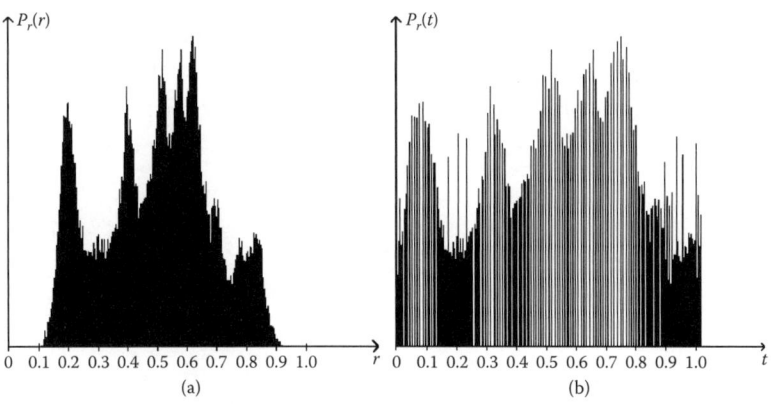

FIGURE 3.3 The histograms before and after equalisation of the image of Lena: (a) the original histogram of Lena, and (b) the histogram after equalisation.

3.1.2.1 Linear Transform

Suppose the maximum and minimum intensities in an image are b_{max} and b_{min}, respectively, and the intensity of each pixel requires B bits of storage, which means that the maximum grey level (intensity) may reach $2^B - 1$. If $0 < b_{min} < b_{max} < 2^B - 1$, one can stretch the pixel intensity as follows:

$$g(x,y) = (2^B - 1) \frac{f(x,y) - b_{min}}{b_{max} - b_{min}} \qquad (3.5)$$

where $f(x,y)$ is the original intensity of the pixel point (x,y), and $g(x,y)$ is the new intensity value after stretching.

3.1.2.2 The Limiting Linear Transform

More generally, one might perform stretching only at the centre of the grey levels, and keep both sides at the given lowest or highest levels. This way, some sensitivity may be reduced. Let τ_1 and τ_2 represent the minimum and maximum thresholds to constrain the stretching part. Assuming that s_1 and s_2 are the possible minimum and maximum intensity values after stretching, respectively, the limiting linear transform can be obtained as follows:

$$g(x,y) = \begin{cases} s_1, & \text{if} \quad f(x,y) \leq \tau_1 \\ \frac{s_2 - s_1}{\tau_2 - \tau_1} \cdot f(x,y) + \frac{s_1 \tau_2 - s_2 \tau_1}{\tau_2 - \tau_1}, & \text{if} \quad \tau_1 < f(x,y) < \tau_2 \quad (3.6) \\ s_2, & \text{if} \quad f(x,y) \geq \tau_2 \end{cases}$$

where $f(x,y)$ is the original intensity of the pixel point (x,y), and $g(x,y)$ is the new intensity value after stretching. Figure 3.4 shows the typical inputs and outputs of the preceding two transformations.

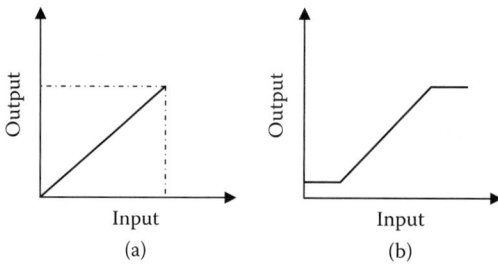

FIGURE 3.4 Typical input and output of (a) a linear transform and (b) a limiting linear transform.

3.2 CONCEPTS AND MODELS OF IMAGE PREPROCESSING

Image processing may be considered as a process of obtaining an output image $g(x,y)$ from a certain input image $f(x,y)$ after a black box operation **P**, that is,

$$g(x,y) = \mathbf{P}[f(x,y)] \tag{3.7}$$

This model [3,4] is depicted in Figure 3.5. It is hoped that, from the input image $f(x,y)$, one can construct and use a system, such as one for image smoothing or image enhancement, to produce the output image $g(x,y)$, which is more suitable for further image processing work and applications. Image smoothing is used to remove any noise, whereas image enhancement is used to enhance certain interesting features in the input image, such as edges, etc.

A clean image $f(x,y)$ may be affected or polluted, and thus it becomes a degraded image in the course of capture, transmission, and storage. As a result, only the degraded image $g(x,y)$ is to be handled, and the original undistorted image $f(x,y)$ is not being dealt with. The degradation may be due to diffraction and image differences in the optics systems, the sensor's nonlinear aberration, the film's nonlinearity, the disturbances due to air turbulences, the spur due to motion of the object, geometric aberration, etc. Using the concept in Equation 3.7, image degrading may also be regarded as applying an unknown system to the original undistorted image. Therefore, the model shown in Figure 3.5 is also applicable to an image-degrading process.

It is often practical to apply, in these processing models, the linear system and the shift-invariant system:

1. The linear system: Given two original images $f_1(x,y)$ and $f_2(x,y)$, the corresponding transformed results $g_1(x,y)$ and $g_2(x,y)$ are obtained by applying a system **P**, that is,

$$g_1(x,y) = \mathbf{P}[f_1(x,y)]$$
$$g_2(x,y) = \mathbf{P}[f_2(x,y)] \tag{3.8}$$

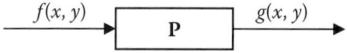

FIGURE 3.5 An image-processing model.

If **P** satisfies

$$\mathbf{P}[\alpha f_1(x,y)+\beta f_2(x,y)]=\alpha \mathbf{P}[f_1(x,y)]+\beta \mathbf{P}[f_2(x,y)]=\alpha g_1(x,y)+\beta g_2(x,y) \quad (3.9)$$

where α and β are constants, **P** is known as a *linear system*.

2. The shift invariant system: Suppose $g(x,y)$ is the transformed result of $f(x,y)$ after applying the process **P**. If **P** satisfies

$$\mathbf{P}[f(x-\alpha, y-\beta)] = g(x-\alpha, y-\beta) \quad (3.10)$$

P is known as a *shift-invariant* system, where α and β are the amount of shifts along the spatial directions. In a shift-invariant system, the transformed result has the same amount of shift as that in the input.

If an image-processing system **P** is a linear shift-invariant (LSI) system, it can be expressed by means of the convolution operation as

$$g(x,y) = \mathbf{P}[f(x,y)] = h(x,y) * f(x,y) = \int_{-\infty}^{+\infty}\int_{-\infty}^{+\infty} h(\alpha,\beta) f(x-\alpha, y-\beta) d\alpha d\beta \quad (3.11)$$

where $h(x, y)$ is known as an *impulse response function* or a *point-spread function*.

As mentioned in Chapter 2, Section 2.1.2, the convolution kernel $h(x, y)$ is usually a symmetric matrix, which is used as a template in the context of image processing. The corresponding discrete convolution formula of Equation 3.11 is given by

$$g(m,n) = h(m,n) * f(m,n) = \sum_{j=-r}^{r}\sum_{k=-s}^{s} h(j,k) f(m+j, n+k) \quad (3.12)$$

The convolution operation in Equation 3.11 is relatively complex. One applies the Fourier transform to images in order to convert the convolution operation in the spatial domain to multiplicative operations in the frequency domain. Let $G(u, v)$, $H(u, v)$, and $F(u, v)$ represent the Fourier transforms of $g(x, y)$, $h(x, y)$, and $f(x, y)$, respectively. The convolution

$$g(x,y) = h(x,y) * f(x,y) \quad (3.13)$$

can be written as the convolution model in the frequency domain as

$$G(u,v) = H(u,v) \cdot F(u,v) \quad (3.14)$$

where $H(u, v)$ is known as the *transfer function*.

3.3 IMAGE SMOOTHING

Noise may be introduced into an image in the process of its construction or transmission. This section is not concerned with how noise is produced; instead, the aim here is to construct systems in order to deal with polluted images, that is, to remove noise to form a smooth image. The main idea of image smoothing is to replace the intensity of every pixel p in an image by a weighted average of the intensities of its neighbouring pixels.

The system used in image smoothing can be either linear or nonlinear. If it is a linear shift-invariant system, one can use convolution to denote it, that is,

$$g(x,y) = h(x,y) * f(x,y) = \int_{-\infty}^{+\infty} \int_{-\infty}^{+\infty} h(\alpha,\beta) f(x-\alpha, y-\beta) d\alpha d\beta \quad (3.15)$$

In essence, constructing a processing system is equivalent to constructing the impulse response function $h(x, y)$ or the convolution kernel $h(m, n)$ for discrete image functions.

3.3.1 Spatial-Domain Methods

Methods applied in the spatial domain include neighbourhood-averaging method, Gaussian filtering, and median filtering. The former two are linear, and hence $h(m, n)$ may be constructed by means of Equation 3.15, followed by convolution.

3.3.1.1 Neighbourhood-Averaging Methods

In neighbourhood-averaging methods, the concept of a processing window is used to define the neighbourhood. The size of a processing window is often chosen as 3×3 or 5×5, which contains the neighbouring pixels surrounding a given pixel. For a 3×3 window, the templates commonly used as the convolution kernel $h(m, n)$ include

$$\begin{bmatrix} 0 & \frac{1}{4} & 0 \\ \frac{1}{4} & 0 & \frac{1}{4} \\ 0 & \frac{1}{4} & 0 \end{bmatrix}, \begin{bmatrix} \frac{1}{8} & \frac{1}{8} & \frac{1}{8} \\ \frac{1}{8} & 0 & \frac{1}{8} \\ \frac{1}{8} & \frac{1}{8} & \frac{1}{8} \end{bmatrix}, \begin{bmatrix} \frac{1}{9} & \frac{1}{9} & \frac{1}{9} \\ \frac{1}{9} & \frac{1}{9} & \frac{1}{9} \\ \frac{1}{9} & \frac{1}{9} & \frac{1}{9} \end{bmatrix}$$

The first one is known as the *4-neighbourhood average*, and the remaining two are *8-neighbourhood averages*. The first two templates do not involve contribution of the intensity of the central pixel, but the third one does. Figure 3.6b depicts the result of a neighbourhood-averaging method.

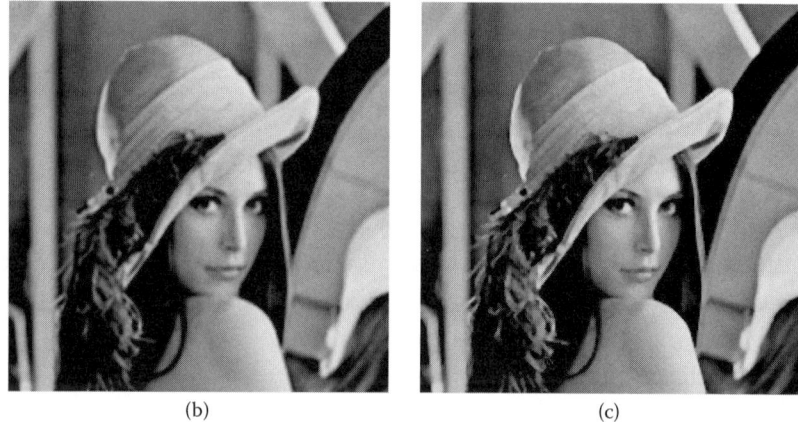

FIGURE 3.6 Images after applying neighbourhood-averaging and median filtering methods: (a) the original image, (b) the resulting image after applying the neighbourhood-averaging method, and (c) the image after median filtering.

Take the 4-neighbourhood average as an example, and consider the pixel located at (i, j). Suppose $f(i, j)$ denotes the original intensity of the pixel located at (i, j), and $g(i, j)$ represents the value after it has been operated on by the first template, one obtains

$$g(i,j) = \frac{1}{4}[f(i-1,j) + f(i+1,j) + f(i,j-1) + f(i,j+1)] \qquad (3.16)$$

3.3.1.2 Threshold-Averaging Methods

The neighbourhood-averaging method is simple and effective but usually leads to image blurring. This is particularly true when the neighbouring pixels are chosen from farther away. To overcome this drawback, a threshold to dispose the averaging result can be adopted. A specific threshold value is chosen in advance and is used to compare with the difference between the resulting intensity and the original intensity. If the difference is greater than the threshold, the original value is replaced by the averaging result; otherwise, the original value is retained.

Consider again the 4-neighbourhood-averaging template, and given the threshold τ, the threshold-averaging result is obtained as follows:

$$a = \frac{1}{4}[f(i-1,j) + f(i+1,j) + f(i,j-1) + f(i,j+1)]$$

$$g(i,j) = \begin{cases} a, & \text{if } |a - f(i,j)| > \tau; \\ f(i,j), & \text{else} \end{cases}$$

(3.17)

3.3.1.3 Gaussian Filtering

Gaussian smoothing filtering [3] is another linear filtering method used in the spatial domain. Its convolution kernel is

$$h(i,j) = e^{-\frac{i^2+j^2}{2\sigma^2}}$$

(3.18)

where σ is a smoothing parameter used to control the extent of smoothing. The larger the value of σ, the greater is the extent of smoothing. Gaussian filtering has many good features, such as rotational symmetry and separability, and has the same smoothing effect in every direction.

In image processing, the Gaussian convolution kernel needs to be transformed to a discrete convolution template. A template with a 7 × 7 window and $\sigma = \sqrt{2}$ is given as follows:

$$\begin{bmatrix} 0.011 & 0.039 & 0.082 & 0.105 & 0.082 & 0.039 & 0.011 \\ 0.039 & 0.135 & 0.287 & 0.368 & 0.287 & 0.135 & 0.039 \\ 0.082 & 0.287 & 0.606 & 0.779 & 0.606 & 0.287 & 0.082 \\ 0.105 & 0.368 & 0.779 & 1.000 & 0.779 & 0.368 & 0.105 \\ 0.082 & 0.287 & 0.606 & 0.779 & 0.606 & 0.287 & 0.082 \\ 0.039 & 0.135 & 0.287 & 0.036 & 0.287 & 0.135 & 0.039 \\ 0.011 & 0.039 & 0.082 & 0.105 & 0.082 & 0.039 & 0.011 \end{bmatrix}$$

For ease of storage and calculation, this template can be converted into an integer form by magnifying each element of the template 91 times, leading to

$$\begin{bmatrix} 1 & 4 & 7 & 10 & 7 & 4 & 1 \\ 4 & 12 & 26 & 33 & 26 & 12 & 4 \\ 7 & 26 & 55 & 71 & 55 & 26 & 7 \\ 10 & 33 & 71 & 91 & 71 & 33 & 10 \\ 7 & 26 & 55 & 71 & 55 & 26 & 7 \\ 4 & 12 & 26 & 33 & 26 & 12 & 1 \\ 1 & 4 & 7 & 10 & 7 & 4 & 1 \end{bmatrix}$$

The sum of the weighted coefficient of the template is $\sum_{i,j} h(i,j) = 1115$. For practical applications, the results should then be divided by 1115.

3.3.1.4 Median Filtering

Median filtering is a nonlinear smoothing method. It sorts the intensities in the neighbourhood window of the reference pixel and calculates the median value of the sorted data. The original value at the reference pixel is then replaced by the median value. Figure 3.6c depicts the result of median filtering.

For simplicity, a one-dimensional signal is used as an example. Suppose the datum at the point i to be dealt with is a_i, and the size of the window is $2k + 1$, where k is an integer. The data in the window is $W_i = (a_{i-k}, a_{i-k+1}, \ldots, a_{i-1}, a_i, a_{i+1}, \ldots, a_{i+k-1}, a_{i+k})$, which is sorted in ascending order as $a_{i_1} \leq a_{i_2} \leq \ldots \leq a_{i_k} \leq a_{i_{k+1}} \leq \ldots \leq a_{i_{2k+1}}$. The median value, which is also the new value to be used at point i, is found to be $a_{i_{k+1}}$.

Example 3.2 Assuming that the one-dimensional discrete signal to be dealt with is given by

{0 0 0 9 0 0 3 4 3 0 2 3 2 0 4 6 4 0 4 7 4 0 0 3 4 5 6 6 6 6 0 0}

let the size of the processing window be 3. Consider the fourth number that has the value 9, which exhibits an impulse noise. The data set in the neighbourhood window is (0,9,0). After sorting, the new data set is (0,0,9). Hence, 0 is used to replace 9. The new signal after processing is

{0 0 0 0 0 0 3 3 3 2 2 2 2 2 4 4 4 4 4 4 4 0 0 3 4 5 6 6 6 6 0 0} ■

In two-dimensional image processing, the size of a processing window is usually chosen as 3 × 3 or 5 × 5. The two-dimensional data is written in a one-dimensional form according to a row-by-row order. The same method described previously may be used to deal with the resulting one-dimensional data.

Example 3.3 The intensities of an image block f are given as

$$f = \begin{bmatrix} 200 & 201 & 202 & 202 & 203 & 202 & 200 & 198 \\ 202 & 203 & 205 & 204 & 204 & 202 & 200 & 197 \\ 205 & 210 & 211 & 212 & 210 & 209 & 208 & 205 \\ 205 & 208 & 213 & 212 & 214 & 210 & 211 & 208 \\ 210 & 212 & 215 & 218 & 217 & 219 & 220 & 218 \\ 212 & 214 & 218 & 220 & 220 & 219 & 218 & 218 \\ 210 & 212 & 213 & 215 & 216 & 216 & 210 & 212 \\ 208 & 208 & 210 & 211 & 212 & 214 & 210 & 210 \end{bmatrix}$$

Consider the pixel point in row three and column three where the intensity is $f(3, 3) = 211$. Its 3 × 3 neighbourhood window is

$$\begin{bmatrix} 203 & 205 & 204 \\ 210 & 211 & 212 \\ 208 & 213 & 212 \end{bmatrix}$$

Rearranging it as one-dimensional data gives

{203 205 204 210 211 212 208 213 212}

After sorting the one-dimensional data, it becomes

{203 204 205 208 210 211 212 212 213}

where the median number is 210. Thus, the new value of the processed image g using median filtering at the position (3, 3) is 210, that is, $g(3, 3) = 210$. ∎

3.3.1.5 Weighted Median Filtering

In median filtering methods, the intensity of each pixel in a window contributes equally to the result. If the intensities of some pixels have more influence to the result, then weighted median filtering [5] should be adopted. Here, "weighted" means increasing the contribution of intensities of some neighbouring pixels, leading to a different median value.

Consider a two-dimensional window of size 3 × 3 and assume that the original intensity of the pixel located at the point (i, j) is $f(i, j)$. The data in its neighbourhood is given by

$$\begin{bmatrix} f(i-1,j-1) & f(i-1,j) & f(i-1,j+1) \\ f(i,j-1) & f(i,j) & f(i,j+1) \\ f(i+1,j-1) & f(i+1,j) & f(i+1,j+1) \end{bmatrix}$$

A template of size 3 × 3 containing weighted values is assigned. The elements of the window are rearranged as a one-dimensional array, following a row–column order in such a way that the intensities in the array repeat according to the corresponding weighting values in the template. The array is then sorted in ascending order, and the median value is used to substitute $f(i, j)$. As an example, suppose the template with the weighted values is given by

$$\begin{bmatrix} 1 & 2 & 1 \\ 2 & 3 & 2 \\ 1 & 2 & 1 \end{bmatrix}$$

Data in the neighbourhood of the pixel (3 × 3) of Example 3.3 now becomes {203, 205, 205, 204, 210, 210, 211, 211, 211, 212, 212, 208, 213, 213, 212}. The series in ascending order becomes {203, 204, 205, 205, 208, 210, 210, 211, 211, 211, 212, 212, 212, 213, 213}, and the median is 211.

3.3.2 Frequency-Domain Methods

By analysing image signals in the frequency domain, one can deal with those high frequencies consisting of components with fast-changing intensities in an image, such as edges, jumps, and grain noise. On the other hand, low frequencies correspond to the slowly varying components of an image, for instance, the background area of the image. Hence, image smoothing is done to filter the high-frequency components and preserve the low-frequency components, which is usually known as *low-pass filtering* [3].

3.3.2.1 Ideal Low-Pass Filtering

According to the image-processing model given in Equation 3.14, image filtering in the frequency domain is completed by constructing the image transfer function $H(u, v)$. The main idea of an ideal low-pass filtering is to preserve the signal with low frequency and cut off the high-frequency signal whose frequency is greater than a preassigned value. Such filtering has the transfer function

$$H(u,v) = \begin{cases} 1, & D(u,v) \leq D_0 \\ 0, & D(u,v) > D_0 \end{cases} \qquad (3.19)$$

where $D(u,v)$ is the distance between the point (u,v) and the origin of the frequency domain:

$$D(u,v) = \sqrt{u^2 + v^2}$$

and D_0 is a specified nonnegative threshold given in advance, called the *cut-off frequency*.

From Equation 3.19, it is found that $H(u,v)$ contains a jump when $D(u,v) = D_0$. This means that its inverse Fourier transform, $h(x,y)$, will be companioned with ringing and blurring phenomena; the result of Equation 3.13, $g(x,y)$, has the same problem. Some improved low-pass filtering methods are introduced to overcome this drawback, such as trapezoidal low-pass filtering and Butterworth low-pass filtering.

3.3.2.2 Trapezoidal Low-Pass Filtering

Trapezoidal low-pass filtering may be used to eliminate the jump mentioned in Section 3.3.2.1. Its transfer function is defined by

$$H(u,v) = \begin{cases} 1, & D(u,v) \leq D_0 \\ \dfrac{D(u,v) - D_1}{D_0 - D_1}, & D_0 < D(u,v) < D_1 \\ 0, & D(u,v) \geq D_1 \end{cases} \qquad (3.20)$$

The meaning of $D(u,v)$ and D_0 is the same as that mentioned in Section 3.3.2.1, and D_1 is a constant satisfying $D_1 > D_0$.

3.3.2.3 Butterworth Low-Pass Filtering

Unlike ideal low-pass filtering, in which the signal is band limited to have only two states, that is, pass or stop, according to the cut-off frequency, Butterworth low-pass filtering makes a slow transition between pass and stop with the transfer function given as follows:

$$H(u,v) = \dfrac{1}{1 + \left[\dfrac{D(u,v)}{D_0}\right]^{2n}} \qquad (3.21)$$

where n is the order of filtering, and D_0 represents the cut-off frequency.

3.4 IMAGE ENHANCEMENT

The image-smoothing methods described in the preceding sections are used to remove noise effects in images. In many situations, it is also important to sharpen some special points and characters of an image, for example, edges and so on, in order to make segmentation and recognition easier. The image-processing model discussed in this section employs different transfer functions to enhance the edges of an image. Such technology is known as the *image-sharpening* process or as *image enhancement*.

The edge of an image represents the fastest varying components of the intensities of an image. It is well known that the gradient operation in calculus may be used to calculate the magnitude and direction of a function in which its function value changes fast. It is possible, in an image function, to make use of the gradient value instead of the original intensity at every pixel in order to preserve the edges of the image.

3.4.1 Gradient

Let $f(x, y)$ denote a grey-scale image function; the gradient at the pixel point (x, y) is defined as the vector

$$\nabla f = \left(\frac{\partial f}{\partial x}, \frac{\partial f}{\partial y} \right) \qquad (3.22)$$

and the direction of the gradient is defined as

$$\theta(x, y) = \arctan\left(\frac{\partial f}{\partial y} \bigg/ \frac{\partial f}{\partial x} \frac{\partial f}{\partial x} \right) \qquad (3.23)$$

where $\theta(x, y)$ denotes the angle between the gradient direction and the x-coordinate. The gradient value, or magnitude of the gradient, for different requirements may be defined by means of the Euclidean norm or l_2-norm:

$$\|\nabla f\| = \sqrt{\left(\frac{\partial f}{\partial x} \right)^2 + \left(\frac{\partial f}{\partial y} \right)^2} \qquad (3.24)$$

l_1-norm:

$$\|\nabla f\| = \left| \frac{\partial f}{\partial x} \right| + \left| \frac{\partial f}{\partial y} \right| \qquad (3.25)$$

or l_∞-norm:

$$\|\nabla f\| = \max\left(\left| \frac{\partial f}{\partial x} \right|, \left| \frac{\partial f}{\partial y} \right| \right) \qquad (3.26)$$

3.4.2 Gradient Image

The gradient image $g(x, y)$ is obtained by replacing the original intensity at every pixel point by its gradient value, that is,

$$g(x,y) = \|\nabla f(x,y)\| \tag{3.27}$$

In practice, some modifications can be applied. For instance, in order to sharpen the edge, it is possible to assign a larger value of intensity nearer to the white colour to the pixel point where its gradient value is greater than a certain threshold, keeping all other points unchanged, that is,

$$g(x,y) = \begin{cases} b_h, & \text{if } \|\nabla f(x,y)\| \geq \tau \\ f(x,y), & \text{else} \end{cases} \tag{3.28}$$

where τ is a preassigned threshold, and b_h is a constant intensity nearer to white. On the other hand, if the location of an edge is of interest, the image may be made as a binary-value image by the use of a certain threshold:

$$g(x,y) = \begin{cases} b_h, & \text{if } \|\nabla f(x,y)\| \geq \tau \\ b_l, & \text{else} \end{cases} \tag{3.29}$$

where τ and b_h are defined as in Equation 3.28, and b_l is a constant intensity nearer to black, which satisfies $b_h > b_l$.

3.4.3 Gradient Operators

Because digital images are defined by means of discrete functions, different methods may be applied to approximate partial derivatives. Various different formulae lead to different possible gradient operators. As an example, consider the locations of the pixels as given in Figure 3.7; one can use the first-order forward difference to approximate the gradient as

$$\frac{\partial f}{\partial x}(i,j) \approx f(i+1,j) - f(i,j)$$
$$\frac{\partial f}{\partial y}(i,j) \approx f(i,j+1) - f(i,j) \tag{3.30}$$

or use the central difference to approximate the gradient as

$$\frac{\partial f}{\partial x}(i,j) \approx \frac{1}{2}[f(i+1,j) - f(i-1,j)]$$
$$\frac{\partial f}{\partial y}(i,j) \approx \frac{1}{2}[f(i,j+1) - f(i,j-1)] \tag{3.31}$$

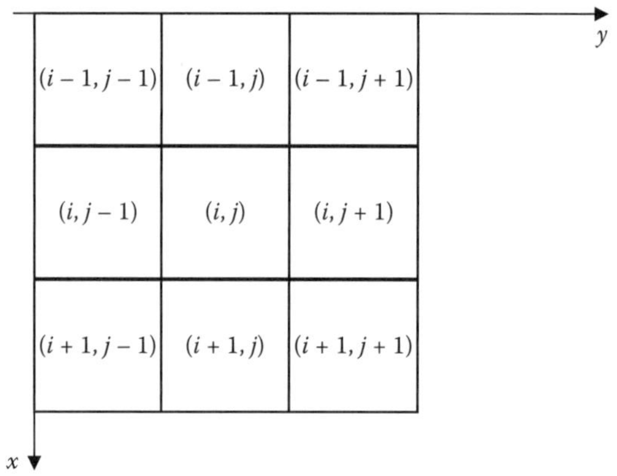

(i − 1, j − 1)	(i − 1, j)	(i − 1, j + 1)
(i, j − 1)	(i, j)	(i, j + 1)
(i + 1, j − 1)	(i + 1, j)	(i + 1, j + 1)

FIGURE 3.7 The position of neighbouring pixels.

Using the image-processing model, Equations 3.30 and 3.31 can be rewritten as convolutions of the image function $f(x, y)$ with the convolution kernels h_x and h_y. In one-dimensional cases, vertical templates representing the processing in the x-coordinate, and horizontal templates representing the processing in y-coordinate are used, and the convolution templates used in the first-order forward difference are

$$h_x = [1,-1]^T;$$
$$h_y = [1,-1]$$
(3.32)

whereas those used in the central difference are

$$h_x = \frac{1}{2}[1,0,-1]^T;$$
$$h_y = \frac{1}{2}[1,0,-1]$$
(3.33)

Thus, Equations 3.30 and 3.31 can be rewritten in a consistent form using convolution as

$$\frac{\partial f}{\partial x} = h_x * f; \qquad \frac{\partial f}{\partial y} = h_y * f \qquad (3.34)$$

The aforementioned difference formulae may introduce large errors because they take into account the intensity of the reference pixel and that of two of its neighbouring pixels only. In fact, the gradient of a pixel ought to be related to every pixel point in its neighbourhood. Some frequently used gradient operators [6] are listed in the following sections.

3.4.3.1 Roberts Operator

The Roberts operator uses l_1-norm to calculate the magnitude of the gradient and adopts the following templates to the partial derivatives $\frac{\partial f}{\partial x}, \frac{\partial f}{\partial y}$:

$$h_x = \begin{bmatrix} 1 & 0 \\ 0 & -1 \end{bmatrix}$$
$$h_y = \begin{bmatrix} 0 & -1 \\ 1 & 0 \end{bmatrix} \quad (3.35)$$

In other words, partial derivatives are calculated using the pixels in the diagonal instead of the pixels in the same row or column. Therefore, one obtains

$$\|\nabla f(i,j)\| = \|f(i+1,j+1) - f(i,j)\| + |f(i,j+1) - f(i+1,j)| \quad (3.36)$$

3.4.3.2 Prewitt Operator

The Prewitt operator adopts Euclidean l_1-norm to compute the gradient values, and $\frac{\partial f}{\partial x}$ and $\frac{\partial f}{\partial y}$ are computed by using the following templates:

$$h_x = \frac{1}{6}\begin{bmatrix} 1 & 1 & 1 \\ 0 & 0 & 0 \\ -1 & -1 & -1 \end{bmatrix}$$
$$h_y = \frac{1}{6}\begin{bmatrix} 1 & 0 & -1 \\ 1 & 0 & -1 \\ 1 & 0 & -1 \end{bmatrix} \quad (3.37)$$

This operator requires six pixel points in the neighbourhood of the reference pixel point. It is in fact using the average of three central differences to approximate the gradient. The effect is better than that of using one central difference formula.

3.4.3.3 Sobel Operator

Similar to the Prewitt operator, the Sobel operator also computes the magnitude of the gradient by using the Euclidean l_2-norm. The difference between these two operators is that, in computing partial derivatives, Sobel considers a higher weighting along the same row or column of the reference pixel point.

$$h_x = \frac{1}{8}\begin{bmatrix} 1 & 2 & 1 \\ 0 & 0 & 0 \\ 1 & -2 & 1 \end{bmatrix}$$

$$h_y = \frac{1}{8}\begin{bmatrix} 1 & 0 & -1 \\ 2 & 0 & -2 \\ 1 & 0 & -1 \end{bmatrix}$$

(3.38)

3.4.3.4 Laplacian Operator

As discussed at the beginning of Section 3.4, an edge has the fastest-changing intensities in an image. By comparing only the gradient value with a preassigned threshold, there may be confusion in identifying the edges. For example, noisy pixel points and nonedge pixel points lying next to an edge can be easily mistaken as edge pixel points. This drawback becomes a disadvantage at the next stage of processing, such as image segmentation and object recognition.

As pointed out by Marr and Hildreth [7], the rapid variation of intensities along an edge corresponds to the local maximum of the derivative at that point. For the second derivative, there are zero-crossing points. It is easier to find the zeros of the second derivative than to calculate the local maximum of the derivative. Hence, using second derivatives, that is, the Laplacian operator,

$$\nabla^2 f = \frac{\partial^2 f}{\partial x^2} + \frac{\partial^2 f}{\partial y^2} \qquad (3.39)$$

plays an important role because it provides the increment of the gradient directly. The following discretised form is often used to calculate the second derivative:

$$\begin{aligned}
\frac{\partial^2 f}{\partial x^2}(i,j) &= \frac{\partial f}{\partial x}\left(\frac{\partial f}{\partial x}(i,j)\right) \\
&\approx \frac{\partial f}{\partial x}(i+1,j) - \frac{\partial f}{\partial x}(i,j) \\
&\approx [f(i+1,j) - f(i,j)] - [f(i,j) - f(i-1,j)] \\
&= f(i+1,j) - 2f(i,j) + f(i-1,j)
\end{aligned} \qquad (3.40)$$

Similarly, one obtains

$$\frac{\partial^2 f}{\partial y^2}(i,j) \approx f(i,j+1) - 2f(i,j) + f(i,j-1) \quad (3.41)$$

Substituting Equations 3.40 and 3.41 in Equation 3.39, one can obtain the convolution template of the Laplacian operator as follows:

$$\nabla^2 = \begin{bmatrix} 0 & 1 & 0 \\ 1 & -4 & 1 \\ 0 & 1 & 0 \end{bmatrix} \quad (3.42)$$

After processing by means of the Laplacian template, one obtains a function that indicates the edge pixel point when it crosses zero.

Figure 3.8 shows the results obtained by using different gradient operators for the same image.

3.4.4 High-Pass Filtering

The methods of image enhancement that use gradient and second derivatives are all convolution-based operations in the spatial domain. Similar to image smoothing, image enhancement can be implemented in the frequency domain. Because the edges of an image usually exhibit high-frequency components in the frequency domain, image enhancement needs to keep the high frequencies and filter the low frequencies. This process is also known as *high-pass filtering* [4].

Ideal high-pass filtering, trapezoidal high-pass filtering, and Butterworth high-pass filtering are commonly used high-pass filtering methods.

3.4.4.1 Ideal High-Pass Filtering

The transfer function of an ideal high-pass filtering is given by

$$H(u,v) = \begin{cases} 0, & D(u,v) \leq D_0 \\ 1, & D(u,v) > D_0 \end{cases} \quad (3.43)$$

where $D(u,v)$ denotes the distance between the point (u,v) and the origin of frequency domain, and D_0 is the cut-off frequency. Here the distance can be calculated as

$$D(u,v) = \sqrt{u^2 + v^2}$$

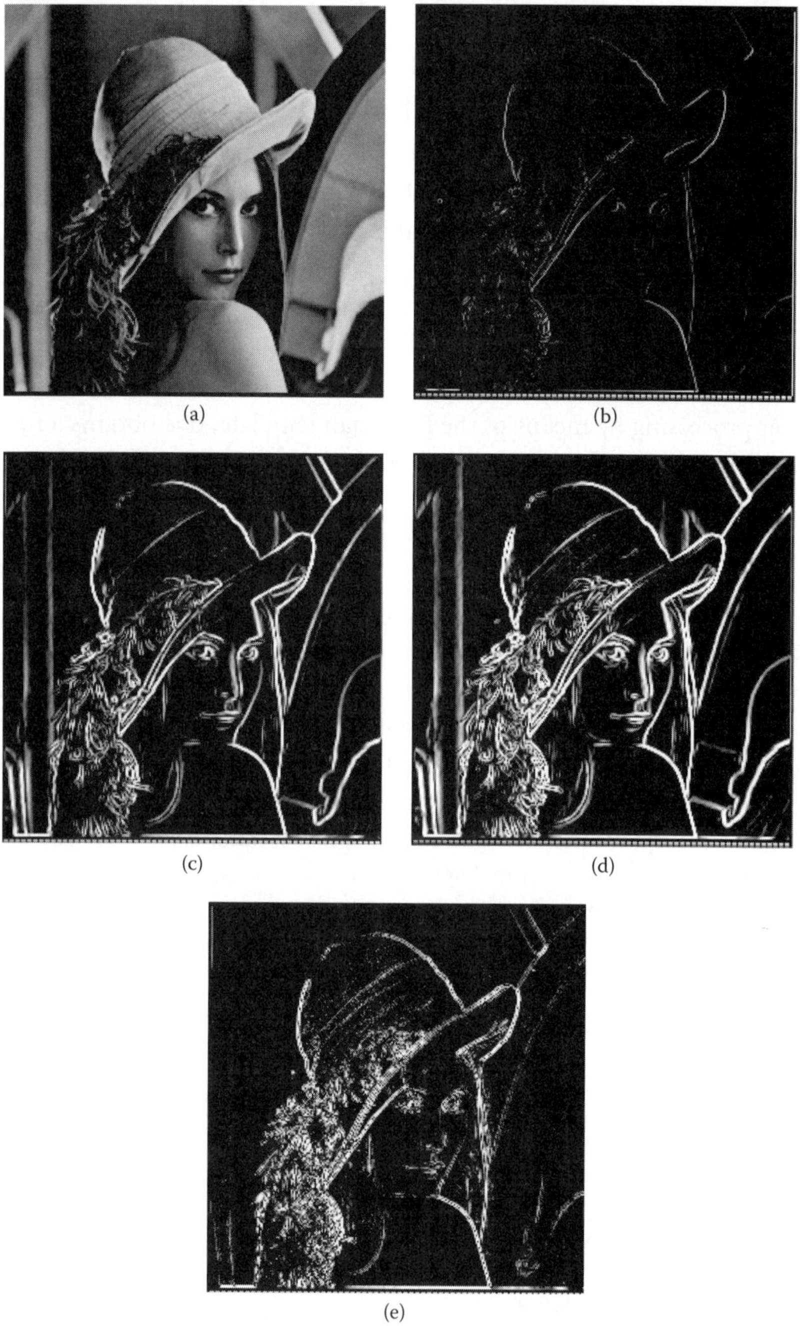

FIGURE 3.8 Results obtained by using different gradient operators: (a) the original image, (b) Roberts operator, (c) Prewitt operator, (d) Sobel operator, and (e) Laplacian operator.

3.4.4.2 Trapezoidal High-Pass Filtering

The trapezoidal high-pass filtering method may be used to eliminate jumps, and the transfer function is given by

$$H(u,v) = \begin{cases} 0, & D(u,v) \leq D_1 \\ \dfrac{D(u,v) - D_1}{D_0 - D_1}, & D_1 < D(u,v) < D_0 \\ 1, & D(u,v) \geq D_0 \end{cases} \quad (3.44)$$

where $D(u, v)$ and D_0 are defined as earlier, and D_1 is a constant satisfying the inequality $D_1 < D_0$.

3.4.4.3 Butterworth High-Pass Filtering

The Butterworth high-pass filtering method makes a slow transition between the pass and the stop. The transfer function is given by

$$H(u,v) = \dfrac{1}{1 + \left[\dfrac{D_0}{D(u,v)}\right]^{2n}} \quad (3.45)$$

where n is the order of filtering, and D_0 is the cut-off frequency.

3.5 IMAGE RESTORATION

3.5.1 Image Degradation Model

Image restoration is the process of recovering an image from its degraded version. A typical restoration method is to construct a degradation model according to prior knowledge of the degradation phenomena. Based on this model, a restoration technique is equivalent to applying an inverse process to the model for restoring the image by satisfying certain criteria.

As described earlier, an image-processing model expressed in Equation 3.7 may be used as a degradation model:

$$g(x,y) = \mathbf{P}[f(x,y)] \quad (3.46)$$

where the input $f(x,y)$ is a clean image, and the output $g(x,y)$ is the degraded image. The operation $\mathbf{P}$ is the degradation system, which may be caused by diffraction and image differences in optics systems, the sensor's nonlinear aberration, the film's nonlinearity, the disturbance due to air turbulence, the spur due to the motion of the object, geometric aberration, etc.

Noise is a common problem in degraded images. Usually, additive noise is preassumed, which is irrelevant to the intensities of images. Gaussian noise and impulse noise are two typical noises. The probability density function $P_G(z)$ of the Gaussian noise is a normal distribution function:

$$P_G(z) = \frac{1}{\sqrt{2\pi}\sigma} e^{-(z-u)/2\sigma^2} \qquad (3.47)$$

where the variable z represents the grey level of a noisy pixel, u is the mathematical expectation of z, and σ is the standard deviation. The probability density function $P_I(z)$ of the impulse noise has the form

$$P_I(z) = \begin{cases} p_a & z = a \\ p_b & z = b \\ 0 & \text{else} \end{cases} \qquad (3.48)$$

where the definition of z is the same as earlier, a and b are two constant grey levels, and p_a and p_b are two constant probability density values. If $a =$ maximum grey level, $b =$ minimum grey level, and $p_a p_b \neq 0$, the impulse noise is also called *salt and pepper noise* [19].

Let $n(x, y)$ denote the noise at the coordinates (x, y). In general, noise may be processed independently by using the degradation system $\mathbf{P}$, and the degradation model can be expressed as

$$g(x, y) = \mathbf{P}[f(x, y)] + n(x, y) \qquad (3.49)$$

Figure 3.9 depicts the image degradation model expressed in this equation. In line with image-processing systems, the image degradation system $\mathbf{P}$ is also assumed to be a linear shift-invariant (LSI) system. By using the convolution operation and the concept of the point-spread function, Equation 3.49 can be expressed as

$$g(x, y) = \mathbf{P}[f(x, y)] + n(x, y) \qquad (3.50)$$
$$= \int_{-\infty}^{+\infty}\int_{-\infty}^{+\infty} h(\alpha, \beta) f(x - \alpha, y - \beta) d\alpha d\beta + n(x, y)$$

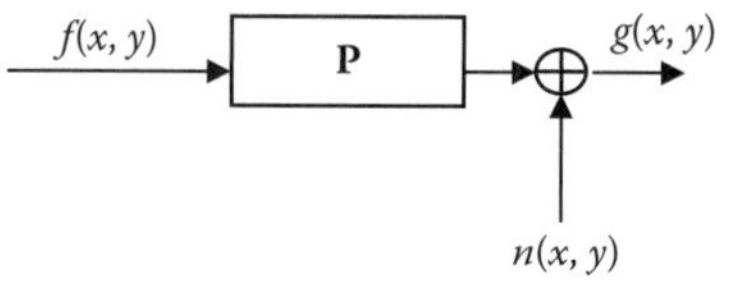

FIGURE 3.9 The image degradation model.

where $f(x,y)$ is the original clean image, $g(x,y)$ is the degraded image, $h(x,y)$ is a point-spread function, and $n(x,y)$ is a certain additive noise function. In discrete notation, the degradation model can be expressed by the discrete convolution

$$g(i,j) = h(i,j) * f(i,j) + n(i,j)$$

$$= \sum_{k=-r}^{r} \sum_{l=-s}^{s} h(k,l) f(i+k, j+l) + n(i,j)$$

where the convolution kernel $h(j,k)$ is assumed to be symmetric with a window size of $(2r+1)$ by $(2s+1)$. Equation 3.49 can also be written in the matrix form

$$\mathbf{g} = \mathbf{Pf} + \mathbf{n} \tag{3.51}$$

where $\mathbf{g}$, $\mathbf{f}$, and $\mathbf{n}$ are vectors corresponding to image functions g and f and the noise function n, and $\mathbf{P}$ is a block circulate matrix.

3.5.2 Image Restoration Based on the Degradation Model

Based on the degradation model mentioned in Section 3.5.1, image restoration can be divided into unconstrained conditional restoration and constrained conditional restoration.

3.5.2.1 Unconstrained Conditional Restoration

The image degradation model expressed in Equation 3.46 does not consider additive noise $n(x,y)$. In this case, according to Equation 3.51, a common restoration method is to choose an approximation vector $\overline{\mathbf{f}}$ of $\mathbf{f}$ and to ensure that the magnitude of $\mathbf{g} - \mathbf{P}\overline{\mathbf{f}}$ is minimised. An objective function can be defined as follows for this purpose:

$$J(\overline{\mathbf{f}}) = \|\mathbf{g} - \mathbf{P}\overline{\mathbf{f}}\|^2 = (\mathbf{g} - \mathbf{P}\overline{\mathbf{f}})^T (\mathbf{g} - \mathbf{P}\overline{\mathbf{f}}) \tag{3.52}$$

$$= \mathbf{g}^T \mathbf{g} - \overline{\mathbf{f}}^T \mathbf{P}^T \mathbf{g} - \mathbf{g}^T \mathbf{P} \overline{\mathbf{f}} + \overline{\mathbf{f}}^T \mathbf{P}^T \mathbf{P} \overline{\mathbf{f}}$$

In order to minimise $J(\overline{\mathbf{f}})$, the derivative of J with respect to $\overline{\mathbf{f}}$ is required to be zero, leading to

$$\frac{\partial J}{\partial \overline{\mathbf{f}}} = -\mathbf{P}^T \mathbf{g} - (\mathbf{g}^T \mathbf{P})^T + 2\mathbf{P}^T \mathbf{P} \overline{\mathbf{f}} = 0 \tag{3.53}$$

$$\overline{\mathbf{f}} = (\mathbf{P}^T \mathbf{P})^{-1} \mathbf{P}^T \mathbf{g} \tag{3.54}$$

When **P** is a square matrix, Equation 3.54 can be simplified as follows:

$$\bar{\mathbf{f}} = (\mathbf{P}^T\mathbf{P})^{-1}\mathbf{P}^T\mathbf{g} = \mathbf{P}^{-1}\mathbf{g} \qquad (3.55)$$

3.5.2.2 Constrained Conditional Restoration

In practice, additive noise is an important component in degraded images and may be predicted a priori. In this situation, a constrained conditional restoration may be used to obtain a reasonable solution. Suppose **Q** is a linear operator operating on an image vector **f**. The problem is to find an optimal image vector $\bar{\mathbf{f}}$ that minimises

$$\|\mathbf{Q}\mathbf{f}\|^2 = (\mathbf{Q}\mathbf{f})^T\mathbf{Q}\mathbf{f} = \mathbf{f}^T\mathbf{Q}^T\mathbf{Q}\mathbf{f} \qquad (3.56)$$

subject to the constraint

$$\|\mathbf{g} - \mathbf{P}\mathbf{f}\|^2 = \|\mathbf{n}\|^2 \qquad (3.57)$$

where **g** is the degraded image vector, and **n** is a given noise vector.

The Lagrange multiplier method may be used to solve this problem. The objective function based on the Lagrange multiplier method is defined as follows:

$$J(\bar{\mathbf{f}}) = \|\mathbf{Q}\bar{\mathbf{f}}\|^2 + \alpha\left(\|\mathbf{g} - \mathbf{P}\bar{\mathbf{f}}\|^2 - \|\mathbf{n}\|^2\right) \qquad (3.58)$$

where α is the Lagrange multiplier. By putting the derivative of J to zero, that is,

$$\frac{\partial J}{\partial \bar{\mathbf{f}}} = 2\mathbf{Q}^T\mathbf{Q}\bar{\mathbf{f}} - 2\alpha\mathbf{P}^T(\mathbf{g} - \mathbf{P}\bar{\mathbf{f}}) = 0 \qquad (3.59)$$

leads to the solution

$$\bar{\mathbf{f}} = \left(\mathbf{P}^T\mathbf{P} + \frac{1}{\alpha}\mathbf{Q}^T\mathbf{Q}\right)^{-1}\mathbf{P}^T\mathbf{g} \qquad (3.60)$$

Both constrained and unconstrained restorations discussed in this section assume that the degradation matrix **P** is known. In practice, **P** is difficult to obtain.

3.5.3 Inverse Filtering

Taking the Fourier transform on both sides of Equation 3.50 leads to

$$G(u,v) = H(u,v)F(u,v) + N(u,v) \quad (3.61)$$

where $G(u, v)$, $H(u, v)$, $F(u, v)$, and $N(u, v)$ represent the Fourier transforms of $g(x, y)$, $h(x, y)$, $f(x, y)$, and $n(x, y)$, respectively. In the absence of noise, Equation 3.61 can be written as

$$G(u,v) = H(u,v)F(u,v) \quad (3.62)$$

If $H(u,v) \neq 0$, Equation 3.62 can be rewritten as

$$F(u,v) = \frac{G(u,v)}{H(u,v)} \quad (3.63)$$

In other words, if the transfer function $H(u, v)$ is known, one can obtain $F(u, v)$ by using Equation 3.63. The original image function can be expressed by means of the inverse Fourier transform

$$f(x,y) = \Gamma^{-1}\left[\frac{G(u,v)}{H(u,v)}\right] \quad (3.64)$$

where the symbol Γ denotes the Fourier transform.

Equation 3.64 does not work in the neighbourhoods of the zeros of $H(u,v)$. Therefore, a processing transfer function $M(u, v)$, defined as follows, is used to avoid this problem [1]:

$$M(u,v) = \begin{cases} \dfrac{1}{H(u,v)} & \sqrt{u^2+v^2} \leq \tau_0 \\ 1 & \sqrt{u^2+v^2} > \tau_0 \end{cases} \quad (3.65)$$

where τ_0 is a specified threshold that satisfies $H(u,v) \neq 0$, when $\sqrt{u^2+v^2} \leq \tau_0$. An approximate function $\hat{F}(u,v)$ of $F(u,v)$ may be obtained by using the following relation:

$$\hat{F}(u,v) = M(u,v)G(u,v) \quad (3.66)$$

Using Equation 3.66, an approximate image function $\hat{f}(x,y)$ can be obtained as

$$\hat{f}(x,y) = \Gamma^{-1}(M(u,v)G(u,v)) \qquad (3.67)$$

This method is known as *inverse filtering*.

When noise exists, in the case of $H(u,v) \neq 0$, Equation 3.61 may be used, leading to

$$F(u,v) = \frac{G(u,v)}{H(u,v)} - \frac{N(u,v)}{H(u,v)} \qquad (3.68)$$

and

$$f(x,y) = \Gamma^{-1}\left[\frac{G(u,v)}{H(u,v)} - \frac{N(u,v)}{H(u,v)}\right] \qquad (3.69)$$

The experimental result shown in Figure 3.10 is obtained when the image has a high value of signal-to-noise ratio (SNR). Note that an inverse restoration cannot produce good results in the case of strong noise. In such situations, one can consider using Wiener filtering.

3.5.4 Wiener Filtering

Consider again the degradation system having noise, expressed in Equation 3.50:

$$g(x,y) = \int_{-\infty}^{+\infty}\int_{-\infty}^{+\infty} h(\alpha,\beta) f(x-\alpha, y-\beta) d\alpha d\beta + n(x,y)$$

(a) The blur image   (b) The result image by using inverse filtering

FIGURE 3.10 Inverse restoration to the cat image, which shows blurring by convolution.

Functions $g(x, y)$, $f(x, y)$, and $n(x, y)$ can be regarded as stable random variables, and $h(x, y)$ is known a priori. The aim of image restoration is to find the best estimation $\hat{f}(x, y)$ of $f(x, y)$ with the minimal mean square error:

$$e^2 = E\{[f(x,y) - \hat{f}(x,y)]^2\} \tag{3.70}$$

where E is the mean operator.

Suppose the approximation $\hat{f}(x, y)$ is written as

$$\hat{f}(x,y) = \int_{-\infty}^{+\infty} \int_{-\infty}^{+\infty} m(x-\alpha, y-\beta) g(\alpha, \beta) d\alpha d\beta \tag{3.71}$$

Then the problem of finding $\hat{f}(x, y)$ is converted into the process of finding $m(x, y)$, which minimises the error function defined by Equation 3.70.

Let $F(u,v)$, $G(u,v)$, $H(u,v)$, and $M(u,v)$ be the Fourier transforms of $f(x,y)$, $g(x,y)$, $h(x,y)$, and $m(x,y)$, respectively. Let $\hat{F}(u, v)$ be the Fourier transform of the approximation $\hat{f}(x, y)$; then $M(u, v)$ and $F(u, v)$ are computed as follows:

$$\hat{F}(u,v) = M(u,v)G(u,v) \tag{3.72}$$

$$M(u,v) = \frac{H^*(u,v)}{|H(u,v)|^2 + P_n(u,v)/P_f(u,v)} \tag{3.73}$$

where $P_n(u, v)$ and $P_f(u, v)$ represent the power spectra of the noisy image and original image, respectively, and $H^*(u, v)$ is the conjugate complex of $H(u, v)$. A detailed discussion may be found in References 3, 8, and 9.

In general, it is very difficult to estimate the power spectra $P_n(u, v)$ and $P_f(u, v)$ accurately. A suitable constant K may be used to approximate $P_n(u, v)/P_f(u, v)$. In the case without noise, that is, $P_n(u, v) = 0$, $M(u, v)$ defined by Equation 3.73 is equivalent to the processing transfer function given by Equation 3.65 in inverse filtering. Figure 3.11 demonstrates the results of an image restoration using Wiener filtering.

3.5.5 Geometric Rectification

As an imaging system itself is nonlinear, or the visual angle is different, geometric distortion may be brought into the image in the process of image shaping. Figure 3.12 depicts several simple examples of geometric distortion.

(a) The noise image   (b) The result image by using Wiener filtering

FIGURE 3.11 Using Wiener filtering to restore the cat image with random noise.

To obtain correct images, one should rectify the image having geometric distortion by using geometric transforms. Geometric rectification involves two steps: the spatial geometric transform of an image, and confirmation of pixel intensities in the rectification space.

3.5.5.1 Spatial Geometric Transforms

A spatial geometric transform rectifies an image $f(x,y)$ having geometric distortion by using the undistorted image $g(u,v)$ or a group of datum marks at which coordinates are known a priori. Figure 3.13 is used to demonstrate a geometric tranform. Using certain reference points known to the two images, one can construct a geometric transform ϕ, describing the relationship between the coordinates (x,y) of the distorted image and the coordinates (u,v) of the undistorted image:

$$\phi = (\phi_1, \phi_2) : (u,v) \to (x,y)$$
$$x = \phi_1(u,v) \quad (3.74)$$
$$y = \phi_2(u,v)$$

Bivariant polynomials may be used to express the geometric transform. Both quadratic and cubic polynomials are able to provide satisfactory results.

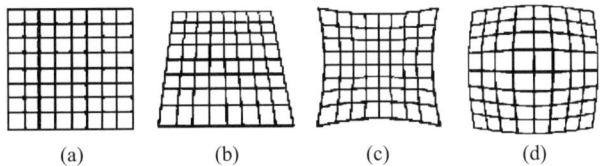

FIGURE 3.12 (a) The original image, (b) the perspective distortion, (c) the pincushion distortion, and (d) the barrel distortion.

Take quadratic polynomials, for example, the geometric transform $\phi = (\phi_1, \phi_2)$ is given by

$$\begin{cases} x = \phi_1(u,v) = a_1 + a_2 u + a_3 v + a_4 u^2 + a_5 uv + a_6 v^2 \\ y = \phi_2(u,v) = b_1 + b_2 u + b_3 v + b_4 u^2 + b_5 uv + b_6 v^2 \end{cases} \qquad (3.75)$$

where a_i and b_i, $1 \leq i \leq 6$, are coefficients to be determined by choosing the corresponding reference points P_j, $0 \leq j \leq M$, with the coordinates (x_j, y_j) in the distorted system and the coordinates (u_j, v_j) in the undistorted system. In order to solve the parameters a_i, $1 \leq i \leq 6$, M should be no less than 6. Using the first part of Equation 3.75, one can obtain the following system of equations for $M = 6$:

$$\begin{bmatrix} x_1 \\ x_2 \\ \vdots \\ x_6 \end{bmatrix} = \begin{bmatrix} 1 & u_1 & v_1 & u_1^2 & u_1 v_1 & v_1^2 \\ 1 & u_2 & v_2 & u_2^2 & u_2 v_2 & v_2^2 \\ & & & & & \\ 1 & u_6 & v_6 & u_6^2 & u_6 v_6 & v_6^2 \end{bmatrix} \begin{bmatrix} a_1 \\ a_2 \\ a_3 \\ a_4 \\ a_5 \\ a_6 \end{bmatrix} \qquad (3.76 \text{ a})$$

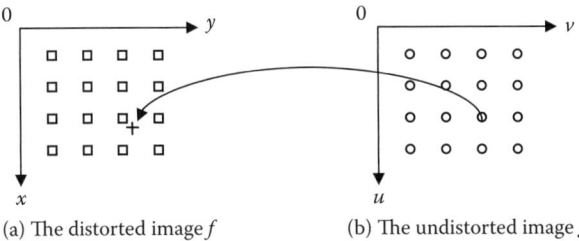

(a) The distorted image f (b) The undistorted image g

FIGURE 3.13 Spatial geometric transform: (a) the distorted image f, and (b) the undistorted image g.

The coefficients $a_1, a_2, ..., a_6$ can be obtained by solving the preceding system of equations. Similarly the coefficients $b_1, b_2, ..., b_6$ can be obtained by solving the following system of equations:

$$\begin{bmatrix} y_1 \\ y_2 \\ \vdots \\ y_6 \end{bmatrix} = \begin{bmatrix} 1 & u_1 & v_1 & u_1^2 & u_1 v_1 & v_1^2 \\ 1 & u_2 & v_2 & u_2^2 & u_2 v_2 & v_2^2 \\ & & & & & \\ 1 & u_6 & v_6 & u_6^2 & u_6 v_6 & v_6^2 \end{bmatrix} \begin{bmatrix} b_1 \\ b_2 \\ b_3 \\ b_4 \\ b_5 \\ b_6 \end{bmatrix} \quad (3.76\ b)$$

In some situations, M may be chosen to be a larger value in order to achieve better rectification. In this case, the least square method may be used to obtain the coefficients $a_1, a_2, ..., a_6$ and $b_1, b_2, ..., b_6$.

3.5.5.2 Confirmation of Pixel Intensities

If the geometric transform maps (u_0, v_0) in the undistorted image to (x_0, y_0) in the distorted image, that is,

$$\phi(u_0, v_0) = (x_0, y_0) \quad (3.77)$$

the intensity of the pixel located at (u_0, v_0) in the undistorted image must be equal to the intensity of the pixel located at (x_0, y_0) in the distorted image, that is,

$$g(u_0, v_0) = f(x_0, y_0) \quad (3.78)$$

However, the values of x_0 and y_0 computed by Equation 3.75 are not necessarily integers. This means that the coordinates (x_0, y_0) might not be pixel coordinates in the distorted image. In this case, $f(x_0, y_0)$ is not defined, and interpolation is required to calculate a value of $f(x_0, y_0)$. There are two methods frequently used in the industry for such interpolation, and are discussed below.

(1) Nearest Interpolation Find a pixel point $(\bar{x}_0, \bar{y}_0)$ in the distorted image, which is the best approximation of (x_0, y_0), and set

$f(x_0, y_0) = f(\bar{x}_0, \bar{y}_0)$. Substitute this value in Equation 3.78, which leads to the following:

$$g(u_0, v_0) = f(\bar{x}_0, \bar{y}_0) \tag{3.79}$$

(2) Bilinear Interpolation This method performs linear interpolation in the two spatial directions by using the intensities of four neighbouring pixels where the coordinates are approximations of (x_0, y_0). Suppose $a = \lfloor x_0 \rfloor$ and $b = \lfloor y_0 \rfloor$, and the four neighbouring pixels are $(a,b), (a+1,b), (a,b+1),$ and $(a+1,b+1)$. The mathematical expression for calculating $f(x_0, y_0)$ by using a bilinear interpolation may be written as

$$f(x_0, y_0) = (1-\alpha)(1-\beta)f(a,b) + (1-\alpha)\beta f(a,b+1) \\ + \alpha(1-\beta)f(a+1,b) + \alpha\beta f(a+1,b+1) \tag{3.80}$$

where $\alpha = x_0 - a$ and $\beta = y_0 - b$.

3.6 PROCESSING METHODS USING PARTIAL DIFFERENTIAL EQUATIONS

From some of the filters discussed in the previous sections, it is natural to link these filters to the finite difference replacement of certain partial differential equations. Hence, it is possible to treat images in a continuous domain that satisfies certain partial differential equations (PDE). In its simplest form, such as Equation 3.42, one can immediately replace the discrete form by means of the Laplace equation in the continuous form. It is then natural to extend and use some of the theories and numerical methods related to solutions of the Laplace equation to produce further convolution templates in handling image processing. On the other hand, it is also natural to extend the use of the Laplace equation to incorporate certain peculiar properties of images.

This chapter describes only two classes of PDE-based image-processing methods: diffusion-based models and total-variation-based models. Readers who are interested in other mathematical models in fluid dynamics should consult References 10 and 11.

3.6.1 Diffusion-Based Models

3.6.1.1 The Heat Conduction Model

In the early part of this chapter, a discrete Gaussian smoothing convolution kernel is given in Equation 3.18. The continuous Gaussian function is defined as follows

$$G_t(x,y) = at^{-1}\exp[-(x^2+y^2)/4t] \tag{3.81}$$

where a is a constant, and t represents the scale parameter. It is possible to show that the convolution of an image function $g(x, y)$ with the Gaussian function

$$I(x,y,t) = G_t(x,y) * g(x,y) \tag{3.82}$$

is equivalent to the steady-state solution of the diffusion equation in two dimensions [12,13]:

$$\begin{cases} \dfrac{\partial I(x,y,t)}{\partial t} = \Delta I(x,y,t) = \dfrac{\partial^2 I}{\partial x^2} + \dfrac{\partial^2 I}{\partial y^2} \in \Omega \\ I(x,y,0) = g(x,y) \end{cases} \tag{3.83}$$

subject to suitable boundary conditions along $\partial \Omega$. Here, Ω denotes the region containing the image, and $\partial \Omega$ its boundary. Equation 3.83 is known as the *isotropic heat conduction model* representing a diffusion process. If the initial image function $g(x,y)$ is noisy, the steady-state solution of this model in Equation 3.83 is a Gaussian smoothing process. Note that edge blurring occurs after the application of Gaussian smoothing as discussed previously.

3.6.1.2 The Anisotropic Diffusion Model

In the isotropic heat conduction model described in Section 3.6.1.1, the diffusion process takes place in the same speed along each direction at a given pixel point of the image, which leads to edge blurring and does not preserve edges during the process of diffusion. It is possible to introduce a nonlinear anisotropic diffusion process [12] governed by the distribution function $c(\|\nabla I\|)$ into the heat conduction model in order to allow the diffusion process to exhibit a maximal speed along the edge direction and to ensure that diffusion terminates along the gradient direction. Taking this discussion into consideration, the mathematical model given by Equation 3.83 may be modified as follows:

$$\begin{cases} \dfrac{\partial I}{\partial t} = div(c(\|\nabla I\|)\nabla I) \in \Omega \\ I(x,y,0) = g(x,y) \end{cases} \qquad (3.84)$$

where $\|\nabla I\| = \sqrt{\left(\dfrac{\partial I}{\partial x}\right)^2 + \left(\dfrac{\partial I}{\partial y}\right)^2}$ is the magnitude of the gradient, and $div = \left(\dfrac{\partial}{\partial x} + \dfrac{\partial}{\partial y}\right)$ represents the divergence of vector. The nonlinear coefficient distribution function $c(x)$ should be designed to preserve the edge, which is also known as the *edge-stopping function*. On the other hand, $c(x)$ needs to be a nonnegative monotonically decreasing function such that $\lim_{x \to \infty} c(x) = 0$. With these concepts in mind, Perona and Malik [12] suggested the coefficient distribution function

$$c(x) = \dfrac{1}{1 + \left(\dfrac{x}{k}\right)^2} \qquad (3.85)$$

where k is a positive constant. It is possible to show that the model (P–M model) obtains better effect only when $xc(x)$ is a nondecreasing function.

The model given by Equation 3.84 is a second-order PDE. Although it can remove noise effectively, experiences of many researchers show that it causes *blocky effect* [12,14]. The reason is that the second-order model replaces the intensities of the neighbourhood of a pixel with a constant intensity, which forms a level horizontal to the x–y plane during an iterative process, leading to the steady-state solution.

To overcome this drawback, fourth-order PDEs are proposed. One common fourth-order PDE model is the Y–K model proposed by You and Kaveh [14], shown as follows:

$$\dfrac{\partial I}{\partial t} = -\nabla^2[c(|\nabla^2 I|)\nabla^2 I] \qquad (3.86)$$

where $c(s)$ is a positive monotonically decreasing function. You and Kaveh [14] take $c(s)$ as

$$c(s) = \dfrac{1}{1 + \left(\dfrac{s}{k}\right)^2} \qquad (3.87)$$

where k is a constant.

Figure 3.14 shows the original Lena image and its version having 10 dB Gaussian noise. Figures 3.15–3.17 show the resulting images obtained by

100 ■ A Concise Introduction to Image Processing Using C++

(a) The original Lena image

(b) Lena image with 10 dB Gaussian noise

FIGURE 3.14 The original Lena image and its version with 10 dB Gaussian noise: (a) the original Lena image, and (b) Lena image with 10 dB Gaussian noise.

using different diffusion models. In particular, Figure 3.15 is obtained using the isotropic diffusion model. Although noise has been smoothed to some extent, the edges of the image become blurry, and this destroys the key features of the original image. From Figure 3.16, one can see that the edges

(a) The restored image by using isotropic diffusion

(b) The top left part of (a)

FIGURE 3.15 Restored images obtained by using isotropic diffusion: (a) the restored image by using isotropic diffusion, and (b) the top left part of (a).

(a) The restored image by using P-M diffusion model

(b) The top left part of (a)

FIGURE 3.16 Restored images obtained by using the P–M diffusion model: (a) the restored image using the P–M diffusion model, and (b) the top left part of (a).

of the image obtained by using the second-order nonlinear diffusion model are preserved. However, the blocky effect is quite obvious. Figure 3.17 illustrates the results obtained by using the fourth-order PDE in overcoming the blocky effect introduced by second-order nonlinear diffusion models.

(a) The restored image by using Y-K diffusion model

(b) The top left part of (a)

FIGURE 3.17 Restored images obtained by using isotropic diffusion: (a) the restored image using the Y–K diffusion model, and (b) the top left part of (a).

3.6.2 TV-Based Models

As described in Section 3.5.2, image restoration can be expressed as a constrained conditional restoration in the continuous domain, with the general image degradation model given by Equation 3.49:

$$g(x,y) = \mathbf{P}[f(x,y)] + n(x,y)$$

where g is the initial noisy image, and f is the final clean image. Image restoration is used to find an optimal image function f, which makes $\|QI\|$ minimum under the constrained condition $\|g - \mathbf{P}I\| = \|n\|$, where Q is a linear operator, and I is the intensity function of an image.

If only noise is considered, the operation $\mathbf{P}$ can be set as an identity operator, that is,

$$\mathbf{P}[f(x,y)] = f(x,y)$$

In a continuous domain, the linear operator Q can be chosen as a functional of gradient [15]:

$$Q(I) = \int_\Omega \varphi(|\nabla I|) dx dy \qquad (3.88)$$

where Ω is the domain of the image I.

Here, a simple case is to choose φ as an identity operator, leading to the total variation (TV) model [16]:

$$\text{Minimise} \int_\Omega \sqrt{\left(\frac{\partial I}{\partial x}\right)^2 + \left(\frac{\partial I}{\partial y}\right)^2} \, dx dy \qquad (3.89)$$

subject to the constraint

$$\int_\Omega \frac{1}{2}(I(x,y) - g(x,y))^2 \, dx dy = \sigma^2 \qquad (3.90)$$

where $\sigma > 0$ is a priori information, representing the standard deviation of the noise function $n(x,y)$. Using the Lagrange multiplier method, the aforementioned TV model can be described as the minimisation of the function $J(I)$ such that

$$J(I) = \int_\Omega \sqrt{\left(\frac{\partial I}{\partial x}\right)I_x^2 + \left(\frac{\partial I}{\partial y}\right)^2}\, dxdy + \frac{1}{2}\lambda \int_\Omega (I-g)^2\, dxdy \qquad (3.91)$$

where λ is the Lagrange multiplier. The solution of the aforementioned minimisation problem can be expressed as a parabolic equation with time as an evolution parameter [16]:

$$\frac{\partial I}{\partial t} = \operatorname{div}\left(\frac{\nabla I}{\|\nabla I\|}\right) - \lambda(I-g), t>0 \text{ and } (x,y) \in \Omega$$

$$I(x,y,0) = g(x,y) \qquad (3.92)$$

$$\frac{\partial I}{\partial n} = 0 \text{ on } \partial\Omega$$

where $I(x,y,t)$ is the image at time t, and n is the outward normal of $\partial\Omega$. When $t \to \infty$, $I(x,y,t)$ approaches a denoised version of $g(x,y)$.

It is easy to find that, when $\lambda = 0$, Equation 3.92 is reduced to a nonlinear diffusion model. In fact, the fourth-order Y–K model given by Equation 3.86 is first expressed as minimising the energy functional [14]:

$$E(I) = \int_\Omega \varphi(|\nabla^2 I|)dxdy \qquad (3.93)$$

where the functional φ satisfies

$$c(s) = \varphi'(s)/s \qquad (3.94)$$

3.6.3 Discrete Formats of PDE Models

The PDEs described in the previous sections are expressed in continuous models. Numerical methods require discretisation along the horizontal, vertical, and temporal axes. Suppose the size of a given image is $N \times N$. Let

$$t = k\Delta t, \quad n = 0,1,2,$$
$$x = ih, \quad i = 0,1,2, \ldots, N-1 \quad (3.95)$$
$$y = jh, \quad j = 0,1,2, \ldots, N-1$$

where h denotes the spatial mesh size, which is usually chosen to be 1, representing the unit distance between two neighbouring pixels, and Δt denotes the temporal step length. In general, an image-processing model based on PDE can always be written as:

$$\frac{\partial I(x,y,t)}{\partial t} = \phi(I(x,y,t)) \quad (3.96)$$

and in a semidiscretised form

$$\frac{I(x,y,t+\Delta t) - I(x,y,t)}{\Delta t} \approx \phi(I(x,y,t)) \quad (3.97)$$

Using the mesh defined in Equation 3.95, it can easily be written in an iterative form

$$I^{k+1}(i,j) = I^k(i,j) + \phi(I^k(i,j)) \cdot \Delta t \quad (3.98)$$

where $I^k(i,j) = I(i,j,k\Delta t)$, and $I^0(i,j) = I(i,j,0)$ is the initial noisy image.

3.7 FURTHER READING

Image preprocessing is an important process because the quality of preprocessing will affect the subsequent processing of images, such as image segmentation or image recognition. It is also a difficult task because it needs a trade-off between noise smoothing and edge preservation. In many cases, different types of noise may occur simultaneously in an image. As a result, numerous works focusing on the study of image preprocessing are currently being carried out. For example, improvements to classical median filtering were proposed, including adaptive median filtering [17,18] and the two-phase denoising method [19]. With the development of the wavelet theory, there are many applications of this theory in image smoothing, and readers may find more details in References 20–22. On the other hand, in recent years, PDE-based image-processing technology is developing rapidly, and applications of the PDE model in image processing have been extended from early image denoising to image segmentation [23,24] and image inpainting [25–28], as well as efficiency improvements in image processing.

3.8 EXERCISES

Q.1 A 3-bit grey-scale image with a size 8×8 is given here. Obtain its grey-scale histogram.

$$f = \begin{bmatrix} 0 & 2 & 2 & 2 & 3 & 2 & 0 & 0 \\ 2 & 3 & 5 & 4 & 4 & 2 & 2 & 1 \\ 5 & 6 & 6 & 7 & 6 & 7 & 6 & 5 \\ 5 & 6 & 7 & 7 & 6 & 4 & 4 & 4 \\ 3 & 3 & 5 & 7 & 7 & 6 & 4 & 3 \\ 2 & 4 & 5 & 6 & 7 & 7 & 6 & 5 \\ 1 & 1 & 2 & 5 & 6 & 6 & 2 & 2 \\ 0 & 2 & 2 & 3 & 4 & 4 & 2 & 0 \end{bmatrix}$$

Q.2 Using Equation 3.5, stretch the image given in Q.1 to 4-bit grey-scale image (the intensity of each pixel in the image is stored in a 4-bit memory).

Q.3 Using the 8-neighbourhood-averaging method (consider the contribution of the intensity of the central pixel) and the 3×3 median filtering method, smooth the image block given in Q1 and compare the results.

Q.4 Add certain amount of Gaussian noise and salt-and-pepper noise to a grey-scale image. Remove the noise using median filtering method, and compare the effects of the two types of noises.

Q.5 Obtain gradient images of the binary image given in Figure 3.18 by using the Prewitt operator and Sobel operator.

Q.6 Assume that an image $f(x,y)$ is the distorted version of a standard image $g(u,v)$. The distortion is the linear transform ϕ, which maps the pixels (2, 0), (0, 3), and (4, 3) in the standard image $g(u,v)$ to the pixels (2, 3), (3, 6), and (4, 4) in the distorted image $f(x,y)$, respectively. Which pixel in the distorted image $f(x,y)$ corresponds to the pixel (1, 5) in the standard image $g(u,v)$, if the nearest neighbouring interpolation is used?

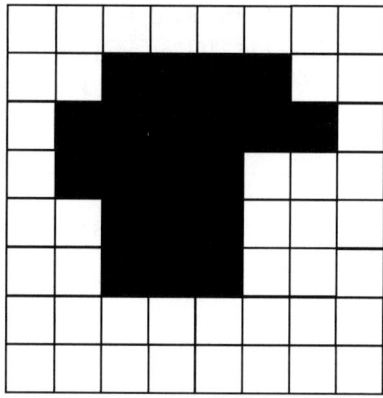

FIGURE 3.18   Q.5

Q.7 Prove that the function given by Equation 3.87 is a positive monotonically decreasing function.

Q.8 Write iterative formulae for the heat conduction model given by Equation 3.83 and the P–M model given by Equation 3.84.

3.9   REFERENCES

1. R. Qiuqi, *Digital Image Processing Science*, Publishing House of Electronic Industry, Beijing, 2001 (in Chinese).
2. M. Sonka, V. Hlavac, and R. Boyle, *Image Processing, Analysis and Machine Vision*, 2nd edition, Thomson Learning and PPTPH, 1998.
3. R. C. Gonzales and R. E. Woods, *Digital Image Processing*, Addison-Wesley, Reading, MA, 1992.
4. K. R. Castleman, *Digital Image Processing*, 2nd Edition, Prentice Hall, Upper Saddle River, NJ, 1996.
5. D. R. K. Brownrigg, The weighted median filter, *Communications of the ACM*, Vol. 27(8): 807–818, 1984.
6. R. Jain, R. Kasturi, and B. G. Schunck, *Machine Vision*, McGraw-Hill, New York, 1995.
7. D. Marr and E. Hildreth, Theory of edge detection, *Proceedings of the Royal Society of London*, Series B, Biological Sciences, Vol. 207: 187–217, 1980.
8. C. W. Helstrom, Image restoration by the method of least squares, *Journal of Optical Society of America*, Vol. 57(3): 297–303, March 1967.
9. W. K. Pratt, Generalised Wiener Filter Computation Techniques, *IEEE Transactions on Computers*, 636–641, July 1972.

10. C. Pozrikidis, *Introduction to Theoretical and Computational Fluid Dynamics*, Oxford University Press, Oxford, U.K., 1996.
11. M. Bertalmio, A. L. Bertozzi, and G. Sapiro, Navier-Stokes, fluid dynamics, and image and video inpainting, *Proceedings of the International Conference on Computer Vision and Pattern Recognition, IEEE*, Vol. I, pp. I-355–I-362, 2001.
12. P. Perona and J. Malik, Scale-space and edge detection using anisotropic diffusion, *IEEE Transactions on Pattern Analysis and Machine Intelligence*, Vol.12(7): 629–639, 1990.
13. S. K. Weeratunga and C. Kamath, PDE-based non-linear diffusion techniques for denoising scientific/industrial images: An empirical study, *Image Processing: Algorithms and Systems, Proceedings of SPIE*, Vol. 4667: 279–290, San Jose, CA, 2002.
14. Y. -L. You and M. Kaveh, Fourth-order partial differential equations for noise removal, *IEEE Transactions on Image Processing*, Vol. 9(10): 1723–1729, 2000.
15. K. Joo and S. Kim, PDE-based image restoration. I: Anti-staircasing and anti-diffusion, Technical report, University of Kentucky, 2003.
16. L. I. Rudin, S. Osher, and E. Fatemi, Nonlinear total variation based removal algorithms, *Physica D*, Vol. 60: 259–268, 1992.
17. H. Hwang and R. A. Haddad, Adaptive median filters: new algorithms and results, *IEEE Transactions on Image Processing*, Vol. 4: 499–502, 1995.
18. S. -C. Hsia, A fast efficient restoration algorithm for high-noise image filtering with adaptive approach, *Journal of Visual Communication and Image Presentation*, Vol. 16(3): 379–392, 2005.
19. R. H. Chan, C. -W. Ho, and M. Nikolova, Salt-and-pepper noise removal by median-type noise detectors and edge-preserving regularization, *IEEE Transactions on Image Processing*, Vol. 14: 1479–1485, 2005.
20. P. R. Coifman and D. L. Donoho, Translation-invariant denoising, In *Wavelets in Statistics of Lecture Notes in Statistics*, A. Antoniadis and G. Oppenheim (Eds.), pp. 125–150, Springer Verlag, New York, 1994.
21. M. Jansen and A. Bultheel, Multiple wavelet threshold estimation by generalized cross validation for images with correlated noise, *IEEE Transactions on Image Processing*, Vol. 8(7): 947–953, 1999.
22. A. Pizurica, W. Philips, I. Lemahieu, and M. Acheroy, A joint inter- and intrascale statistical model for Bayesian wavelet based image denoising, *IEEE Transactions on Image Processing*, Vol. 11(5): 545–557, 2002.
23. X. Du, D. Cho, and T. D. Bui, Image inpainting and segmentation using hierarchical level set method, *Proceedings of the 3rd Canadian Conference on Computer and Robot Vision*, IEEE Computer Society, 2006.

24. X. -C. Tai, O. Christiansen, P. Lin, and I. Skjaelaaen, Image segmentation using some piecewise constant level set methods with MBO type of project, *International Journal of Computer Vision,* Vol. 73(1): 61–76, 2007.
25. M. Bertalmio, G. Sapiro, V. Caselles, and C. Ballester, Image inpainting, In *Proceedings SIGGRAPH 2000, Computer Graphics Proceedings,* Kurt Akeley (Ed.), pp. 417–424, Addison-Wesley, Reading, MA, 2000.
26. M. Bertalmio, A. L. Bertozzi, and G. Sapiro, Navier-Stokes, Fluid Dynamics, and Image and Video Inpainting, In *Proceedings ICCV 2001,* pp. 1335–1362, IEEE Computer Society Press, New York, 2001.
27. T. F. Chan and J. Shen, Non-texture inpainting by curvature driven diffusions (CDD), *Journal of Visual Communication and Image Representation,* Vol. 12(4): 436–449, 2001.
28. T. F. Chan and J. Shen, Mathematical models for local non-texture inpaintings, *SIAM Journal on Applied Mathematics,* Vol. 62(3): 1019–1043, 2001.

3.10 PARTIAL CODE EXAMPLES

Project 3-1: Show the Grey-Scale Histogram of an 8-Bit Grey-Scale Image

(These codes can be found in CD: Project3-1\source code\DlgHistShow.cpp)

```
#include "stdafx.h"
#include "project3_1.h"
#include "DlgHistShow.h"
#include "project3_1Doc.h"
#ifdef _DEBUG
#define new DEBUG_NEW
#undef THIS_FILE
static char THIS_FILE[] = __FILE__;
#endif
BOOL CDlgHistShow::OnInitDialog()
{
    CDialog::OnInitDialog();

    // TODO: Add extra initialization here

    // Set the pointer pointing to the pixel intensity
of original image
    unsigned char * lpSrc;
    // cyclic variable
    int i,j;
```

```
    // get the histogram-show dialogue item
    CWnd* pWnd = GetDlgItem(IDC_DLG_HIST_SHOW);
    int cxDIB = (int) ::DIBWidth(lpDIB); // Size of
DIB - x
    int cyDIB = (int) ::DIBHeight(lpDIB); // Size of
DIB - y
    LPSTR lpDIBBits=::FindDIBBits (lpDIB);
    // count the number of byte of the image per line
    long lLineBytes = WIDTHBYTES(cxDIB * 8);
    // reset the counter to 0
    for (i = 0; i < 256; i ++)
    {
        //
        m_nHist[i] = 0;
    }

    // compute the pixel number of each grey scale and
get the histogram
    for (i = 0; i < cyDIB; i ++)
    {
        for (j = 0; j < cxDIB; j ++)
        {
            // the pointer pointing to the i-th
line and j-th column picture pixel
            lpSrc = (unsigned char*)lpDIBBits +
lLineBytes * (cyDIB - 1 - i) + j;

            // add 1 to the counter
            m_nHist[*(lpSrc)]++;
        }
    }
    return TRUE; // return TRUE unless you set the
focus to a control
    // EXCEPTION: OCX Property Pages should return
FALSE
}
void CDlgHistShow::OnPaint()
{
    CPaintDC dc(this); // device context for painting
```

```cpp
// TODO: Add your message handler code here
// cyclic variable
int i;
// get the histogram-show dialogue item
CWnd* pWnd = GetDlgItem(IDC_DLG_HIST_SHOW);
// get the context of the DIB
CDC* pDC = pWnd->GetDC();
pWnd->Invalidate();
pWnd->UpdateWindow();
pDC->Rectangle(0, 0, 330, 300);
// create the object of the Pen
CPen* pPenRed = new CPen;
// create red pen (draw the axis of the coordinates)
pPenRed->CreatePen(PS_SOLID, 1, RGB(255,0,0));
// select the red pen and save the previous pen
CPen* pOldPen = pDC->SelectObject(pPenRed);
// draw the axis
pDC->MoveTo(10,10);

// draw the Y-axis
pDC->LineTo(10, 280);
// draw the X-axis
pDC->LineTo(320, 280);
// draw the scales in X-axis
CString strTemp;
strTemp.Format("0");
pDC->TextOut(10, 283, strTemp);
strTemp.Format("50");
pDC->TextOut(60, 283, strTemp);
strTemp.Format("100");
pDC->TextOut(110, 283, strTemp);
strTemp.Format("150");
pDC->TextOut(160, 283, strTemp);
strTemp.Format("200");
pDC->TextOut(210, 283, strTemp);
strTemp.Format("255");
pDC->TextOut(265, 283, strTemp);

//
```

```
for (i = 0; i < 256; i += 5)
{
        if ((i & 1) == 0)
        {
                // the times of 10
                pDC->MoveTo(i + 10, 280);
                pDC->LineTo(i + 10, 284);
        }
        else
        {
                // the times of 5
                pDC->MoveTo(i + 10, 280);
                pDC->LineTo(i + 10, 282);
        }
}

// draw the arrowhead of the X-axis
pDC->MoveTo(315,275);
pDC->LineTo(320,280);
pDC->LineTo(315,285);

// draw the arrowhead of the Y-axis
pDC->MoveTo(10,10);
pDC->LineTo(5,15);
pDC->MoveTo(10,10);
pDC->LineTo(15,15);
// the maximum counter in the histogram
LONG lMaxCount = 0;
// compute the maximum counter
for (i = 0; i <= 255; i ++)
{
        //
        if (m_nHist[i] > lMaxCount)
        {
                // update the maximum counter
                lMaxCount = m_nHist[i];
        }
}
```

```
        // output the maximum counter
        pDC->MoveTo(10, 25);
        pDC->LineTo(14, 25);
        strTemp.Format("%d", lMaxCount);
        pDC->TextOut(11, 26, strTemp);

        // create a new pen object
        CPen* pPenBlue = new CPen;
        // create a blue pen (draw the histogram)
        pPenBlue->CreatePen(PS_SOLID, 1, RGB(0,0,255));
        // select the blue pen
        pDC->SelectObject(pPenBlue);
        // decide whether the maximum counter exist
        if(lMaxCount > 0){
                // draw the histogram
                for (i = 0; i <= 255; i ++)
                {
                        pDC->MoveTo(i + 10, 280);
                        pDC->LineTo(i + 10, 281 - (int) (m_
nHist[i] * 256 / lMaxCount));
                }
        }
        // restore the previous pen
        pDC->SelectObject(pOldPen);

        // delete the red pen and the blue pen
        delete pPenRed;
        delete pPenBlue;
}
```

Project 3-2: Median Filtering

(These codes can be found in CD: Project3-2\source code\project3_2View.cpp)

```
#include "stdafx.h"
#include "project3_2.h"
#include "math.h"
#include "project3_2Doc.h"
#include "project3_2View.h"
#ifdef _DEBUG
```

```
#define new DEBUG_NEW
#undef THIS_FILE
static char THIS_FILE[] = __FILE__;
#endif
/***********************************************************
*****************
*
* Function name:
* GetMedianValue()
*
* parameters:
* unsigned char * pUnchFltValue      - the pointer
pointing to the array which needs to decide the median
*  int iFilterLen                - the length of the array
*
* return value:
* unsigned char                - return the median of
the array。
*
* Description:
* The function uses bubble sort method to rearrange the
array data in order and
* return the median value.
*
***********************************************************
****************/
unsigned char GetMedianValue(unsigned char * pUnchFlt-
Value, int iFilterLen)
{
        // cyclic variables
        int          i;
        int          j;

        // temp variable
        unsigned char bTemp;

        // rearrange the data in the array in order using
the bubble sort method
            for (j = 0; j < iFilterLen - 1; j ++)
```

```
        {
                for (i = 0; i < iFilterLen - j - 1; i ++)
                {
                        if (pUnchFltValue[i] > pUnchFltValue[i + 1])
                        {
                                // swap
                                bTemp = pUnchFltValue[i];
                                pUnchFltValue[i] = pUnchFltValue[i + 1];
                                pUnchFltValue[i + 1] = bTemp;
                        }
                }
        }

        // compute the median
        if ((iFilterLen & 1) > 0)
        {
                // return the median value if the number of the array is odd
                bTemp = pUnchFltValue[(iFilterLen + 1) / 2];
        }
        else
        {
        // return the average of the two median values if the number of the array is even
                bTemp = (pUnchFltValue[iFilterLen / 2] + pUnchFltValue[iFilterLen / 2 + 1]) / 2;
        }

        // return the median
        return bTemp;
}
/***********************************************************
 *
 * function name:
 * MedianFilter()
```

```
*
* \input parameters:
* LPSTR lpDIB            - information of the original
image
* int nTempWidth         - the width of the template
* int nTempHeight        - the height of the template
* int nTempCenX          - the X-coordinate of the cen-
tre of the template
* int nTempCenY          - the Y-coordinate of the cen-
tre of the template
*
* \ return value:
* BOOL                   - return TRUE if success other-
wise return FALSE
*
* Description:
* The function performs the median filtering for the
given image.
*
***************************************************
****************/
BOOL MedianFilter(LPSTR lpDIB,int nTempWidth, int nTem-
pHeight,
                                int nTempCenX, int
nTempCenY)
{
        // the pointer pointing to the temporary image data
        LPBYTE lpImage;

        // cyclic variables
        int i, j, k, l;

        // the pointer pointing to the original image
        unsigned char*      lpSrc;

        // the pointer pointing to the region which will
be copied
        unsigned char*      lpDst;
```

```cpp
    LPSTR lpDIBBits=::FindDIBBits (lpDIB);
    int cxDIB = (int) ::DIBWidth(lpDIB); // Size of
DIB - x
    int cyDIB = (int) ::DIBHeight(lpDIB); // Size of
DIB - y
    long lLineBytes = WIDTHBYTES(cxDIB * 8); // count
the the number of byte of the image per line
    // the pointer pointing to the filter array
    unsigned char* pUnchFltValue;
    // allocate the memory for the temp data
    lpImage = (LPBYTE) new char[cxDIB*cyDIB];
    // decide success or not
    if (lpImage == NULL)
    {
        // return
        return FALSE;
    }

    // copy the original image data to the temp data
memory
    memcpy(lpImage, lpDIBBits, cxDIB*cyDIB);

    // allocate temp memory to save the filter array
    pUnchFltValue = new unsigned char[nTempHeight *
nTempWidth];

    // decide success or not
    if (pUnchFltValue == NULL)
    {
        // release the allocated memory
        delete[] lpImage;

        // return
        return FALSE;
    }

    // median filtering
    // row
```

Preprocessing Techniques for Images ■ 117

```
        for(i = nTempCenY; i < cyDIB - nTempHeight + nTempCenY + 1; i++)
        {
                // column
                for(j = nTempCenX; j < cxDIB - nTempWidth + nTempCenX + 1; j++)
                {
                        // points to the data of the ith row, jth column of the new DIB
                        lpDst = (unsigned char*)lpImage + cxDIB * (cyDIB - 1 - i) + j;

                        // read the filter array
                        for (k = 0; k < nTempHeight; k++)
                        {
                                for (l = 0; l < nTempWidth; l++)
                                {
                                        // points to the data of the (i - nTempCenY + k) row,
                                        // (j - nTempCenX + l) column of DIB
                                        lpSrc = (unsigned char*) lpDIBBits + cxDIB * (cyDIB - 1 - i + nTempCenY - k) + j - nTempCenX + l;

                                        // save the intensity of the pixel
                                        pUnchFltValue[k * nTempWidth + l] = *lpSrc;
                                }
                        }

                        // get the median
                        * lpDst = GetMedianValue(pUnchFltValue, nTempHeight * nTempWidth);
                }
        }

        // copy the result image
```

```
            memcpy(lpDIBBits, lpImage, cxDIB*cyDIB);

            // release the memory
            delete[]lpImage;
            delete[]pUnchFltValue;
            // return
            return TRUE;
}
/*****************
```

Project 3-3: Gradient Image Obtained by Using Sobel Operator

(These codes can be found in CD: Project3-3\source code\project3_3View.cpp)

```
#include "stdafx.h"
#include "project3_3.h"
#include "math.h"
#include "project3_3Doc.h"
#include "project3_3View.h"
#ifdef _DEBUG
#define new DEBUG_NEW
#undef THIS_FILE
static char THIS_FILE[] = __FILE__;
#endif
/***********************************************************
*****************
  *
  * \function name:
  * SobelOperator()
  *
  * \input parameters:
  * LPSTR lpDIB            - information of the original
image
  * double * pdGrad    - the pointer pointing to the infor-
mation of the gradient
  * image
  *
  * \return value:
  * null
  *
```

```
 * \Description:
 * Sobel operator
 *
 *
 ***********************************************************
 *****************
 */
void SobelOperator(LPSTR lpDIB, double * pdGrad)
{
            // go through the y-coordinate of the pixel
of the original image
        int y;

        // go through the x-coordinate of the pixel of the
original image
        int x;

 // the pointer pointing to the data of the original image
        unsigned char *lpSrc;
        LPSTR lpDIBBits=::FindDIBBits (lpDIB);
        int cxDIB = (int) ::DIBWidth(lpDIB); // Size of
DIB - x
        int cyDIB = (int) ::DIBHeight(lpDIB); // Size of
DIB - y
        long lLineBytes = WIDTHBYTES(cxDIB * 8); // count
the the number of byte of the image per line
        // the width and the height of the image
        int nWidth                  = cxDIB                     ;
        int nHeight                 = cyDIB                     ;

        // initialisation
        for(y=0; y<nHeight ; y++)
              for(x=0 ; x<nWidth ; x++)
              {
                      *(pdGrad+y*nWidth+x)=0;
              }

              // set the weights of the template
              static int nWeight[2][3][3]  ;
```

```
nWeight[0][0][0] = -1 ;
nWeight[0][0][1] = 0 ;
nWeight[0][0][2] = 1 ;
nWeight[0][1][0] = -2 ;
nWeight[0][1][1] = 0 ;
nWeight[0][1][2] = 2 ;
nWeight[0][2][0] = -1 ;
nWeight[0][2][1] = 0 ;
nWeight[0][2][2] = 1 ;

nWeight[1][0][0] = 1 ;
nWeight[1][0][1] = 2 ;
nWeight[1][0][2] = 1 ;
nWeight[1][1][0] = 0 ;
nWeight[1][1][1] = 0 ;
nWeight[1][1][2] = 0 ;
nWeight[1][2][0] = -1 ;
nWeight[1][2][1] = -2 ;
nWeight[1][2][2] = -1 ;



//supporting window
int nTmp[3][3];

// temp variables
double dGrad ;
double dGradOne;
double dGradTwo;

// cyclic variables for the template
int yy ;
int xx ;


// compute the gradient magnitude of each pixel in the original image
// by using Sobel operator
//
```

```
                for (y=1; y < nHeight-1; y++)
                    for (x=1; x < nWidth-1; x++)
                    {
                        lpSrc = (unsigned char*)lpDIB-Bits;

                        dGrad = 0 ;
                        dGradOne = 0 ;
                        dGradTwo = 0 ;
                        // the intensities of the supporting window

                        // the first row
                        nTmp[0][0] = lpSrc[(y-1)*cxDIB + x - 1] ;

                        nTmp[0][1] = lpSrc[(y-1)*cxDIB + x] ;

                        nTmp[0][2] = lpSrc[(y-1)*cxDIB + x + 1] ;

                        // the second row
                        nTmp[1][0] = lpSrc[y*cxDIB + x - 1] ;

                        nTmp[1][1] = lpSrc[y*cxDIB + x] ;

                        nTmp[1][2] = lpSrc[y*cxDIB + x + 1] ;

                        // the third row
                        nTmp[2][0] = lpSrc[(y+1)*cxDIB + x - 1] ;

                        nTmp[2][1] = lpSrc[(y+1)*cxDIB + x] ;

                        nTmp[2][2] = lpSrc[(y+1)*cxDIB + x + 1] ;

                        // gradient magnitude
                        for(yy=0; yy<3; yy++)
                            for(xx=0; xx<3; xx++)
                            {
```

```
                                        dGradOne +=
nTmp[yy][xx] * nWeight[0][yy][xx] ;
                                        dGradTwo +=
nTmp[yy][xx] * nWeight[1][yy][xx] ;
                              }
                              dGrad = dGradOne*dGradOne
+ dGradTwo*dGradTwo ;
                              dGrad = sqrt(dGrad) ;
                              // save the gradient
magnitude to the memory

*(pdGrad+y*nWidth+x)=dGrad;
                    }
}
/***********************************************
*****************
*
* \ function name:
* OnEdgeSobel()
*
* \ input parameter:
* null
*
* \ return value:
* null
*
* \ Description:
* image segmentation using Sobel operator
*
************************************************
****************
*/
void CProject3_3View::OnEdgeSobel()
{
        // change the style of the cursor
        BeginWaitCursor();
        // cyclic variables
        int x;
        int y;
```

```
        CProject3_3Doc * pDoc = (CProject3_3Doc *)this-
>GetDocument();
        ASSERT_VALID(pDoc);
        if(pDoc->m_hDIB == NULL)
                return ;
        LPSTR lpDIB = (LPSTR) ::GlobalLock((HGLOBAL)
pDoc->m_hDIB);
        LPSTR lpDIBBits=::FindDIBBits (lpDIB);
        int cxDIB = (int) ::DIBWidth(lpDIB); // Size of
DIB - x
        int cyDIB = (int) ::DIBHeight(lpDIB); // Size of
DIB - y
        long lLineBytes = WIDTHBYTES(cxDIB * 8);
 // count the number of byte of the image per line
        // the pointer pointing to the gradient data
        double * pdGrad;

        // allocate the memory for the gradient image data
        pdGrad=new double[cxDIB*cyDIB];

        //the pointer pointing to the image data
        unsigned char *lpSrc;

        // apply Sobel operator to compute the gradient
magnitude for each pixel
        SobelOperator(lpDIB, pdGrad);

 //thresholding the gradient image
        for(y=0; y<cyDIB ; y++)
             for(x=0 ; x<cxDIB ; x++)
             {
                     lpSrc = (unsigned char*)lpDIBBits;
                     if(*(pdGrad+y*cxDIB+x)>50)
                          *(      lpSrc+y*cxDIB+x
)=BYTE(0);
                     else
                          *(      lpSrc+y*cxDIB+x
)=BYTE(255);
             }
```

```
    // release the memory
        delete pdGrad;
        pdGrad=NULL;
        // restore the style of the cursor
        EndWaitCursor();

        // set the modification flag
        pDoc->SetModifiedFlag(TRUE);

        // update the view
        pDoc->UpdateAllViews(NULL);
}
/
*************************************************************
****************
```

Project 3-4: Image Restoration Using the Second- and Fourth-Order Partial Differential Equations

(These codes can be found in CD: Project3-4\source code\imageprocess-Dlg.cpp)

```
#include "stdafx.h"
#include "project3_4.h"
#include "imageprocessDlg.h"
#include "io.h"
#include "math.h"
#include "stdlib.h"
#include "stdio.h"
#include <fcntl.h>
#include <errno.h>
#include <sys/types.h>
#include <sys/stat.h>
#ifdef _DEBUG
#define new DEBUG_NEW
#undef THIS_FILE
static char THIS_FILE[] = __FILE__;
#endif
void CimageprocessDlg::OnProcess()
{
if(b) // use the second-order PDE model
```

```
{
        int i,j;
        int count=0, t; //iteration times
CString msg;
        double k;
        double l=0.25;

        UpdateData(true);
        if (m_done)
        {
                CDialog::OnOK();
                return;
        }

        // the noisy image
        m_noise_image = (unsigned char **)
malloc((m_height)
* sizeof (unsigned char *));
row_image = (unsigned char *)malloc((long)(m_
width)*(long)(m_height)*sizeof(unsigned char));
        if (row_image == NULL)
        {
                Message("Error: Out of memory from
image buffer");
                return;
        }
        for (i = 0; i<m_height; ++i, row_image +=
m_width)
                m_noise_image[i] = row_image;

        // read the noisy image
        if(!ReadNoiseImage(m_width, m_height))
                return;

        // the smoothing image, the initial value is
the same as the noisy image
        m_smooth_image = (double **)malloc((m_
height)*sizeof(double *));
```

```
                smooth_row_image = (double *)malloc((long)
(m_width)*(long)(m_height)*sizeof(double));
                if (smooth_row_image == NULL)
                {
                        Message("Error: Out of memory from
image buffer");
                        return;
                }
                for (i = 0; i<m_height; ++i, smooth_row_
image += m_width)
                        m_smooth_image[i] = smooth_row_image;

                for( i=0; i<m_height; ++i)
                        for(j=0; j<m_width; ++j)
                                m_smooth_image[i][j] = (double)
m_noise_image[i][j];

                // temporary image
                m_temp_image = (double **)malloc(
(m_height+2)*sizeof(double *));
temp_row_image = (double *)malloc((long)(m_
width+2)*(long)(m_height+2)*sizeof(double));
                if(temp_row_image == NULL)
                {
                        Message("Error: Out of memory
from image buffer");
                        return;
                }
                for(i=0; i<m_height+2; ++i,temp_row_
image += (m_width+2))
                        m_temp_image[i] = temp_row_
image;

                for(i=0;i<m_height+2;++i)
                        for(j=0;j<m_width+2;j++)
                                m_temp_image[i][j] = 0;

                        for(i=1; i<m_height+1; ++i)
                                for(j=1; j<m_width+1; ++j)
```

```
                            {
                                    m_temp_image[i][j] 
= (double) m_noise_image[i-1][j-1];
                            }
                            // perform the process 
iteratively
                            for (t=0; t < m_itera-
tions; t++)
                            {
                                    ++count;
                                    //m_temp_image for 
the temporary image
                                    for (j=1; j<m_
width+1; j++)
                                    {
                                            m_temp_
image[m_height+1][j] 
= (double) m_temp_image[m_height][j];
                                            m_temp_
image[0][j] = (double) m_temp_image[1][j];
                                    }
                                    for(i=1;i<m_
height+1;i++)
                                    {
                                            m_temp_
image[i][m_width+1] = 
(double) m_temp_image[i][m_width];
                                            m_temp_
image[i][0] = (double) m_temp_image[i][1];
                                    }
                                    // maximum itera-
tion times
                                    k = 10;
                                    //p-m, isotropic 
diffusion, minimum surfaces
                                    for(i=0;i<m_
height;i++)
                                            for(j=0;j<m_
width;j++)
```

```
                                                    m_
smooth_image[i][j] = m_temp_image[i+1][j+1]
                                                  +1*(
gg(m_temp_image[i+2][j+1]-m_temp_image[i+1][j+1],k)*(m_
temp_image[i+2][j+1]-m_temp_image[i+1][j+1])
                                                +gg(m_
temp_image[i][j+1]-m_temp_image[i+1][j+1],k)*(m_temp_
image[i][j+1]-m_temp_image[i+1][j+1])
                                                +gg(m_
temp_image[i+1][j+2]-m_temp_image[i+1][j+1],k)*(m_temp_
image[i+1][j+2]-m_temp_image[i+1][j+1])
                                                +gg(m_
temp_image[i+1][j]-m_temp_image[i+1][j+1],k)*(m_temp_
image[i+1][j]-m_temp_image[i+1][j+1]));

                                        for(i=1;i<m_
height+1;i++)

for(j=1;j<m_width+1;j++)

m_temp_image[i][j] = m_smooth_image[i-1][j-1];


                                                  msg.
Format("interations %d.", count);

Message(msg);
                                }
                                msg.Format("interations
%d.", count);
                                Message(msg);
                                // Write smooth image to
file
                                if(!WriteSmoothImage(m_
width, m_height))
                                        return;

                                // release memory
                                free(m_noise_image[0]);
```

```
                    free(m_temp_image[0]);
                    free(m_smooth_image[0]);

                    m_ProcessButton.SetWindowText("Close");
                    m_done=true;
            }
    else // use the fourth-order PDE model
    {
            int i,j;
            int count=0, k; // iteration times
            CString msg;

            UpdateData(true);
            if (m_done)
                    {
                            CDialog::OnOK();
                            return;
                    }
                    // the noisy image
m_noise_image = (unsigned char **)malloc((m_height)*size of (unsigned char *));
            row_image = (unsigned char *)malloc((long)(m_width)*
(long)(m_height)*size of(unsigned char));
                    if (row_image == NULL)
                            {
                                    Message("Error: Out of memory from image buffer");
                                    return;
                            }
                    for (i = 0; i<m_height; ++i, row_image += m_width)
                                    m_noise_image[i] = row_image;

                    // read the noisy image
                    if(!ReadNoiseImage(m_width, m_height))
                                    return;
```

/ the smoothed image, the initial value is the same as the noisy image
```
                    m_smooth_image = (double **)
malloc((m_height)*size of(double *));
                    smooth_row_image = (double *)
malloc((long)(m_width)*
(long)(m_height)*size of(double));
                    if (smooth_row_image == NULL)
                        {
                                Message("Error:
Out of memory from image buffer");
                                return;
                        }
                    for (i = 0; i<m_height; ++i, smooth_
row_image += m_width)
                                m_smooth_image[i]
= smooth_row_image;


                    for(i=0; i<m_height; ++i)
                        for(j=0; j<m_width; ++j)
                                            m_smooth_
image[i][j] = (double)m_noise_image[i][j];


                    // temporary image u
                    m_u_image = (double **)malloc((m_
height+2)*size of(double *));
                    u_row_image = (double *)malloc((long)
(m_width+2)*
(long)(m_height+2)*sizeof(double));
                    if (u_row_image == NULL)
                        {
                                Message("Error: Out of
memory from image buffer");
                                return;
                        }
                    for (i = 0; i<m_height+2; ++i, u_row_
image += (m_width+2))
                                        m_u_image[i]
= u_row_image;
```

```
                for (i = 1; i<=m_height; ++i)
                    for (j=1;j<=m_width;j++)
                        m_u_image[i][j] = (double)m_noise_image[i-1][j-1];

                //the divergence of u
                m_grads_u = (double **)malloc((m_height)*sizeof(double *));
                grads_u_row_image = (double *)malloc((long)(m_width)*
                (long)(m_height)*sizeof(double));
                if (grads_u_row_image == NULL)
                {
                    Message("Error: Out of memory from image buffer");
                    return;
                }
                for (i = 0; i<m_height; ++i, grads_u_row_image += m_width)
                                                                    m_grads_u[i] = grads_u_row_image;

                //the coefficient function g
                g = (double **)malloc((m_height+2)*sizeof(double *));
                g_row_image = (double *)malloc((long)(m_width+2)*
                (long)(m_height+2)*sizeof(double));
                if (g_row_image == NULL)
                {
                    Message("Error: Out of memory from image buffer");
                    return;
                }
                for (i = 0; i<m_height+2; ++i, g_row_image += (m_width+2))
                                                                    g[i] = g_row_image;
```

```
                // the divergence of the function g
                m_grads_g = (double **)malloc((m_height)*size of(double *));
                grads_g_row_image = (double *) malloc((long)(m_width)*
(long)(m_height)*size of(double));
                if (grads_g_row_image == NULL)
                {
                    Message("Error: Out of memory from image buffer");
                    return;
                }
                for (i = 0; i<m_height; ++i, grads_g_row_image += m_width)
                                                                    m_grads_g[i] = grads_g_row_image;

                //perform the process iteratively
                for (k=0; k<m_iterations; k++)
                {
                    ++count;
                    //initialize the border of u
                    for (i=1; i<=m_height; i++)
                                                            m_u_image[i][0] = m_u_image[i][1];
                    for (i=1; i<=m_height; i++)
m_u_image[i][m_width+1] = m_u_image[i][m_width];
                    for (j=1;j<=m_width;j++)
                                                            m_u_image[0][j] = m_u_image[1][j];
                    for (j=1;j<=m_width;j++)
m_u_image[m_height+1][j] = m_u_image[m_height][j];

                    // divergence of u
                    for (i=0; i<m_height; i++)
                        for(j=0; j<m_width; j++)
```

Preprocessing Techniques for Images ■ 133

```
                                        m_grads_u[i]
[j] = grads_u_function(i,j);


                                // g (div (u))
                                for (i=1; i<=m_height; i++)
                                    for (j=1; j<=m_width; j++)
                                                        g[i][j]= g_function(i,j);


                                // the border
                                for (i=1; i<=m_height; i++)
                                                  g[i][0] = g[i][1];
                                for (i=1; i<=m_height; i++)
                                                  g[i][m_width+1] = g[i][m_width];
                                for (j=1; j<=m_width; j++)
                                                  g[0][j] = g[1][j];
                                for (j=1; j<=m_width; j++)
                                                  g[m_height+1][j] = g[m_height][j];


                                // divergence of g
                                for (i=0; i<m_height; i++)
                                    for (j=0; j<m_width; j++)
m_grads_g[i][j] = grads_g_function(i,j);


                                // the result image of iteration
                                for (i=0; i<m_height; i++)
                                    for(j=0; j<m_width; j++)
                m_smooth_image[i][j] =
```

```
                    (-m_t_step)*m_grads_g[i][j]+m_u_image[i+1][j+1];


                                    for(i=1; i<=m_height; i++)
                                        for(j=1; j<=m_width; j++)
                                            m_u_image[i][j] = m_smooth_image[i-1][j-1];


                                    msg.Format("interations %d.", count);
                                    Message(msg);
                                }
                                msg.Format("interations %d.", count);
                                Message(msg);

                                // Write the smoothed image to a file
                                if (!WriteSmoothImage(m_width, m_height))
                                    return;

                                // release memory
                                free(m_noise_image[0]);
                                free(m_u_image[0]);
                                free(m_grads_u[0]);
                                free(g[0]);
                                free(m_grads_g[0]);
                                free(m_smooth_image[0]);

                                m_ProcessButton.SetWindowText("Close");
                                m_done=true;
                            }
                        }
```

CHAPTER 4

Image Segmentation

Image segmentation is used to distinguish interesting objects from an image. Objects are considered to be basic elements used in image analysis and image understanding. For the convenience of description, several important concepts related to images are briefly described here [1]. A region Ω of an image is a set of pixels adjacent to each other. Any two pixel points in the region are connected by a path that itself consists of a number of pixels. An object of interest may be embedded in a region. Some such objects of interest may be considered as foreground, whereas others are considered as background of the image. The border $\partial\Omega$ is a set of pixels forming the boundary of Ω. For any pixel on $\partial\Omega$, there is at least one neighbouring pixel that is outside the region Ω. The concepts of regions, objects, and borders concern the positions of pixels. An edge $e(i, j)$ at the pixel point (i, j) of an image with the image function $f(x, y)$ is defined by means of a vector with two components, the magnitude $M(i, j)$, and the direction $\theta(i, j)$, as follows:

$$M(i,j) = \|\nabla f(i,j)\|$$

$$\theta(i,j) = \arg\tan\left(\frac{\frac{\partial f}{\partial y}(i,j)}{\frac{\partial f}{\partial x}(i,j)}\right) - \frac{\pi}{2}$$

The edge magnitude $M(i, j)$ is the magnitude of the gradient, and the edge direction $\theta(i, j)$ is rotated 90° clockwise with respect to the gradient direction. Edges are very useful in finding the border of a region.

Thresholding, edge-based segmentation, and region-based segmentation are common segmentation technologies. The first section of this chapter is devoted to a discussion on thresholding using a number of different concepts, leading to an optimal thresholding algorithm, followed by a section on edge-based segmentation, and another on region-based segmentation.

4.1 THRESHOLDING

A simple method of segmentation is to use a threshold to partition an image into two parts, namely, foreground and background. The threshold τ is used to check against the intensity $f(i, j)$ of a pixel at point (i, j) of the image. If the background of the image is dark, then the foreground consists of bright objects:

$$g(i,j) = \begin{cases} 1, & f(i,j) \geq \tau \\ 0 & f(i,j) < \tau \end{cases} \tag{4.1}$$

In other words, the pixel at (i, j) of the image f belongs to the foreground if $g(i, j) = 1$; otherwise, it belongs to the background. If one is interested in dark objects with a light background, Equation 4.1 is rewritten as

$$g(i,j) = \begin{cases} 1, & f(i,j) < \tau \\ 0 & f(i,j) \geq \tau \end{cases} \tag{4.2}$$

This thresholding method is used to partition an image according to the intensities at different pixels of the image. In practice, thresholding can also be applied to other properties of images, such as colour, texture, gradient, etc.

4.1.1 Semi-Thresholding and Band–Thresholding

There are many modifications [1] to the basic thresholding based on Equations 4.1. and 4.2. Semithresholding and band thresholding are two typical modifications based on Equation 4.1, and their definitions are given here. Figure 4.1 depicts the results of four basic methods of thresholding segmentation.

4.1.1.1 Semi-Thresholding

$$g(i,j) = \begin{cases} f(i,j), & f(i,j) \geq \tau \\ 0 & f(i,j) < \tau \end{cases} \tag{4.3}$$

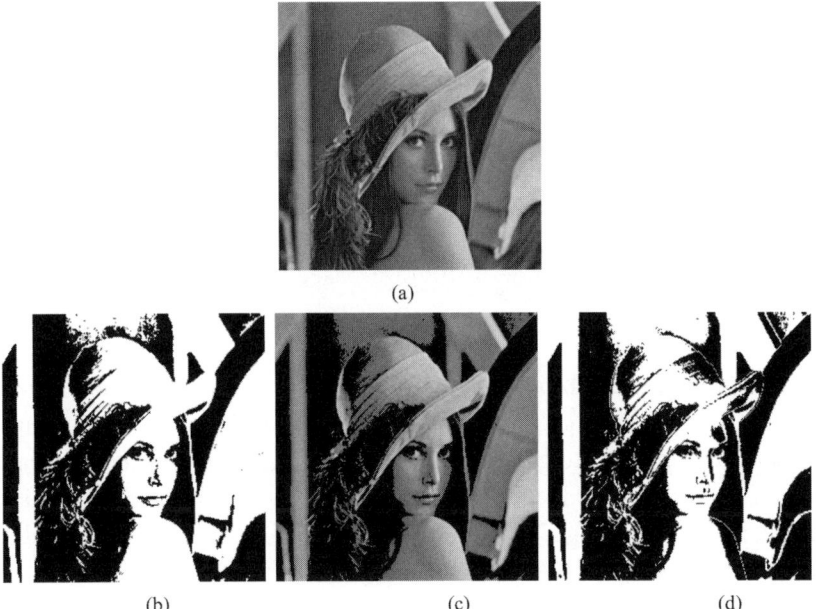

FIGURE 4.1 Basic thresholding segmentation: (a) original image, (b) basic thresholding, (c) semi-thresholding, and (d) band-thresholding.

This modification of thresholding is called *semi-thresholding*; it keeps the foreground unchanged and puts the background as black.

4.1.1.2 Band-Thresholding
In this modification of thresholding, a threshold interval U defined by two thresholding values is used instead of a single value τ.

$$g(i,j) = \begin{cases} 1, & f(i,j) \in U \\ 0 & f(i,j) \notin U \end{cases} \quad (4.4)$$

A pixel (i, j) belongs to the foreground if its intensity belongs to U.

4.1.2 Histogram-Based Thresholding

The crucial problem of thresholding is how to choose a proper threshold τ for an image. In most cases, the choice of thresholds is based on grey-level histograms of images, as introduced in Chapter 3.

4.1.2.1 The Mode Method

If an image contains similar grey-level objects that vary from the grey levels of the background, its grey-level histogram consists of two peaks [1,2], one belonging to the objects and the other belonging to the background. Such a histogram is called a *bi-model*. Usually, the valley between the two peaks—a minimum histogram value that corresponds to the minimal number of pixels in the grey level—is selected as a threshold. Figure 4.2 shows the result of the mode-histogram-based thresholding.

4.1.2.2 Adaptive (Local) Method

In some images, brightness may be nonuniform over the whole image, or the distribution of grey levels of the background may be nonuniform. These situations lead to grey-level histograms with more than two peaks, and the mode method does not work. A better way to handle such situations is to partition the image into several subimages and obtain a threshold for each subimage using the mode method [3].

4.1.3 Optimal (Iterative) Thresholding

An alternative method to the histogram-based method is optimal thresholding, which is obtained by means of an iterative process. For a simple version of this method [4], an initial threshold is set to roughly partition the image into foreground and background. The mean value of the average intensity of the foreground and the average intensity of the background is calculated and used as a better approximate threshold. The iteration continues until no new values of the threshold can be obtained. Figure 4.3 shows a typical example of applying optimal thresholding segmentation. It can be described using the following algorithm:

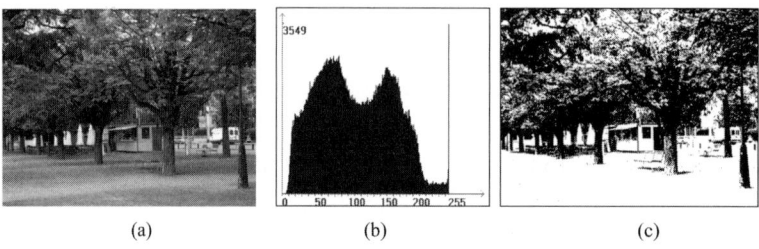

FIGURE 4.2 The mode-histogram-based thresholding: (a) original image, (b) its grey-level histogram with threshold = 120, and (c) histogram-based thresholding.

(a)   (b)

FIGURE 4.3 optimal thresholding: (a) original image, and (b) optimal thresholding segmentation (optimal threshold = 115).

Algorithm 4.1: Optimal thresholding segmentation

For the given image $f(i, j)$: $0 \leq i \leq m - 1$, $0 \leq j \leq n - 1$

Set an initial threshold $\tau^{(0)}$, e.g., $\tau^{(0)} = \dfrac{1}{n \times m} \sum_{i,j} f(i,j)$;

Set $k = 0$; $\delta = 0.0001$;

Do

{

    Partition the pixels of f into two sets:

$$\Omega_1 = \{(i,j): f(i,j) \geq \tau^{(k)}\};$$

$$\Omega_2 = \{(i,j): f(i,j) < \tau^{(k)}\}$$

    Compute the average intensities of two sets:

$$\mu_1 = \dfrac{1}{|\Omega_1|} \sum_{(i,j) \in \Omega_1} f(i,j)$$

$$\mu_2 = \dfrac{1}{|\Omega_2|} \sum_{(i,j) \in \Omega_2} f(i,j)$$

where $|\Omega_1|$ and $|\Omega_2|$ denote the number of pixels in Ω_1 and Ω_2, respectively.

    Compute the new threshold:

$$\tau^{(k+1)} = \dfrac{1}{2}(\mu_1 + \mu_2);$$

$$k := k + 1;$$

} until { $|\tau^{(k)} - \tau^{(k-1)}| < \delta$ }

End-Algorithm

4.2 EDGE-BASED SEGMENTATION

This section introduces methods of finding the border of an object and identifying the object or the foreground of an image. These methods are usually referred to as edge-based segmentation, and include gradient processing and border tracing.

4.2.1 Edge Image Thresholding

In Chapter 3, gradient operators, such as those introduced by Roberts, Prewitt, and Sobel, were discussed with applications for edge enhancement of an image. The application of thresholding to the results obtained by means of a gradient operator may be used to identify the border of an image (See Chapter 3, Sections 3.4.2 and 3.4.3). However, such methods, which calculate the gradient magnitude image with thresholding, will broaden the edge of the image, and hence, will affect the accurate location of the edge. On the other hand, any edges retrieved can be easily corrupted by noise. Canny [5, 6] proposed a multiple detection method that avoids these two shortcomings. It involves two steps. First, the image is smoothed by means of a Gaussian filter in order to reduce the effect of noise. Second, the gradient direction of the gradient image is processed with the non-maximal suppression method and is used for edge thinning.

Given the original image $f(i, j)$, $0 \leq i \leq m-1$, $0 \leq j \leq n-1$, and by using the notation for neighbouring pixels shown in Figure 4.4, the process of Canny's edge detection algorithm consists of the following four steps:

1. Use a Gaussian filter h to smooth the image f leading to the smoothed result $S = h * f$. The template of the Gaussian smoothing function h can be found in Chapter 3, Section 3.3.1.3.

2. Compute the gradient magnitude $G(i, j)$ and the gradient direction $\theta(i, j)$ of the pixel at point (i, j) of the image function S by using the formulae

$$G(i,j) = \sqrt{(\frac{\partial S}{\partial x}(i,j))^2 + (\frac{\partial S}{\partial y}(i,j))^2}$$

$$\theta(i,j) = \arctan\left(\frac{\frac{\partial S}{\partial y}(i,j)}{\frac{\partial S}{\partial x}(i,j)}\right) \quad (4.5)$$

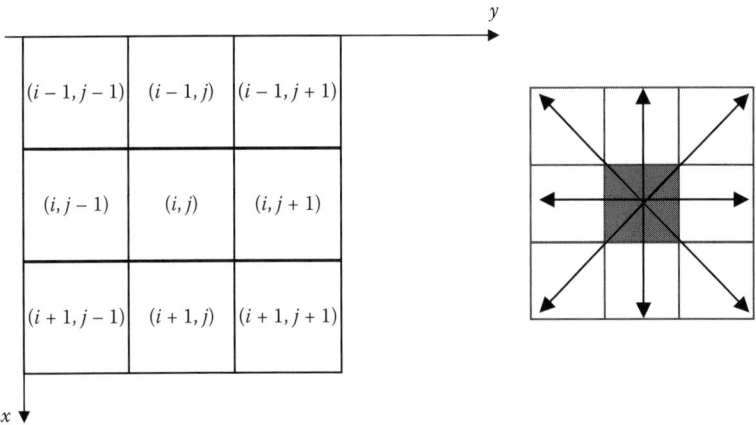

FIGURE 4.4 The neighbourhood of the pixel at point and the gradient directions.

where the approximate discrete formulae of the two partial derivatives are given by

$$\frac{\partial S}{\partial x}(i,j) = \frac{1}{2}[S(i+1,j) - S(i,j) + S(i+1,j+1) - S(i,j+1)]$$
$$\frac{\partial S}{\partial y}(i,j) = \frac{1}{2}[S(i,j+1) - S(i,j) + S(i+1,j+1) - S(i+1,j)]$$
(4.6)

3. Determine the edge pixel using nonmaximal suppression. The characteristic of an edge pixel is that its gradient magnitude is the local maximal in the gradient direction. To determine whether the gradient magnitude of the pixel at (i, j) is a local maximal or not, one needs to locate the two neighbouring pixels p_1 and p_2 of the pixel at point (i, j) and calculate the gradient magnitudes of the three pixels. Suppose pixels p_1 and p_2 are located at positions (i_1, j_1) and (i_2, j_2), respectively. If the gradient magnitude of the pixel at position (i, j) is maximum, it is an edge point, and the gradient magnitude is used as its intensity; otherwise, the pixel is not an edge point, and its intensity is set to 0. The resulting image φ can be described as follows:

$$\varphi(i,j) = \begin{cases} G(i,j), & \text{if} \quad G(i,j) \geq G(i_1,j_1) \quad \text{and} \quad G(i,j) \geq G(i_2,j_2) \\ 0, & \text{otherwise} \end{cases}$$
(4.7)

Because the locations of pixels are discrete, gradient directions also need to be quantized. Take the 8-neighbouring domain, as shown in Figure 4.4, with the pixel at position (i, j) as an example. The positions (i_1, j_1) and (i_2, j_2) of the neighbouring pixels p_1 and p_2 in the gradient direction can be computed as follows:

a. If $-\frac{1}{8}\pi < \theta(i,j) \leq \frac{1}{8}\pi$, $\theta(i,j)$ is quantized as 0, and

$$(i_1, j_1) = (i, j-1), (i_2, j_2) = (i, j+1);$$

b. If $\frac{1}{8}\pi < \theta(i,j) \leq \frac{3}{8}\pi$, $\theta(i,j)$ is quantized as $\frac{1}{4}\pi$, and

$$(i_1, j_1) = (i+1, j-1), (i_2, j_2) = (i-1, j+1);$$

c. If $-\frac{3}{8}\pi < \theta(i,j) \leq -\frac{1}{8}\pi$, $\theta(i,j)$ is quantized as $-\frac{1}{4}\pi$, and

$$(i_1, j_1) = (i-1, j-1), (i_2, j_2) = (i+1, j+1);$$

d. If $\frac{3}{8}\pi < \theta(i,j) < \frac{1}{2}\pi$ or $-\frac{1}{2}\pi < \theta(i,j) < -\frac{3}{8}\pi$, $\theta(i,j)$ is quantized as $\frac{\pi}{2}$, and

$$(i_1, j_1) = (i-1, j), (i_2, j_2) = (i+1, j).$$

4. Thresholding with hysteresis. Non-maximal suppression reduces the border of an object to the width of just one pixel. Due to the existence of noise and thin texture, this process may result in spurious responses, which lead to streaking problem. *Streaking* means the breaking up of an edge contour caused by the operator fluctuating above and below the threshold. Hysteresis using two thresholds $\tau_1 < \tau_2$ can eliminate streaking. If the value $\varphi(i, j)$ of the pixel at position (i, j) in the resulting image is larger than τ_2, the pixel is definitely an edge pixel, and all such edge pixels constitute the edge output. Any pixel connected to this edge pixel and has its value larger than τ_1 is selected as an edge pixel. The following algorithm details the thresholding with hysteresis.

Algorithm 4.2: Thresholding with hysteresis

Let $\Omega(i, j)$ denote the 8-neighbourhood of the pixel at (i, j).

For the given image $\varphi(i, j) : 0 \leq i \leq m-1, 0 \leq j \leq n-1$

Prepare two thresholds: $\tau_1 < \tau_2$;
Initialize the resulting edge image $E(i,j) = 0 : 0 \leq i \leq m-1, 0 \leq j \leq n-1$;
Repeat
{   count = 0;
          For $i = 0$ to $m - 1$
          For $j = 0$ to $n - 1$
          Do
                    If $(\varphi(i,j) \geq \tau_2)$ then { $E(i,j) = 1$; count = count +1}
                    else if $(\varphi(i,j) \geq \tau_1)$ then for each $(k, 1) \in \Omega(i,j)$
                              if $E(k, l) = 1$ then { $E(i,j) = 1$; count = count +1}
          End-Do
} until (count = 0)
Output the resulting edge image: $E(i, j)$.
End-Algorithm.
Figure 4.5 shows the result of an edge image detected by the Canny method.

4.2.2 Edge Relaxation

Certain parts of the border resulting from previous processing are often missed due to noise. The missing parts lead to disconnected borders. Edge relaxation is a processing method similar to thresholding with hysteresis. The pixels between two sets of border pixels are considered border pixels only if their neighbouring pixels are taken into account and criteria is used to determine whether border pixels are relaxed [7,8].

(a) Original Lena image   (b) Edge image detected by Canny method

FIGURE 4.5 Edge detection using the Canny method: (a) Original Lena image, and (b) Edge image detected by Canny method.

A classic edge relaxation method is based on the concept of crack edges [9]. There are four crack edges attached to the pixel at (i, j) that are defined by its relation to its 4-neighbours:

$e_\rightarrow(i,j)$:

$$M(e_\rightarrow(i,j)) = |f(i,j+1) - f(i,j)|$$

$$\theta(e_\rightarrow(i,j)) = \begin{cases} 0, & \text{if } f(i,j+1) \geq f(i,j) \\ \pi, & \text{if } f(i,j+1) < f(i,j) \end{cases}$$

$e_\uparrow(i,j)$:

$$M(e_\uparrow(i,j)) = |f(i-1,j) - f(i,j)|$$

$$\theta(e_\uparrow(i,j)) = \begin{cases} \dfrac{\pi}{2}, & \text{if } f(i-1,j) \geq f(i,j) \\ \dfrac{3}{2}\pi, & \text{if } f(i-1,j) < f(i,j) \end{cases}$$

$e_\leftarrow(i,j)$:

$$M(e_\leftarrow(i,j)) = |f(i,j-1) - f(i,j)|$$

$$\theta(e_\leftarrow(i,j)) = \begin{cases} \pi, & \text{if } f(i,j-1) \geq f(i,j) \\ 0, & \text{if } f(i,j-1) < f(i,j) \end{cases}$$

$e_\downarrow(i,j)$:

$$M(e_\downarrow(i,j)) = |f(i+1,j) - f(i,j)|$$

$$\theta(e_\downarrow(i,j)) = \begin{cases} \dfrac{3\pi}{2}, & \text{if } f(i+1,j) \geq f(i,j) \\ \dfrac{1}{2}\pi, & \text{if } f(i+1,j) \geq f(i,j) \end{cases}$$

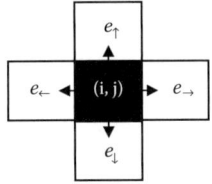

FIGURE 4.6 Directions of the four crack edges attached to a pixel.

The directions of the four crack edges are depicted in Figure 4.6. The pixel positions related to the crack edges are called *end vertices*. For example, the end vertices of the crack edge $e_\rightarrow(i,j)$ are (i,j) and $(i, j+1)$, and the end vertices of the crack edge $e_\uparrow(i,j)$ are (i,j) and $(i-1, j)$.

The main idea of edge relaxation is to decide whether a crack edge can be used to extend a continuous border based on the properties of its neighbours. Each crack edge e is assigned a confidence, representing its strength of being a border part. The initial confidence $c^{(0)}(e)$ may be taken as its normalised magnitude, with normalisation based on either the global maximum of the crack edges in the entire image or on a local maximum in some large neighbourhood of the edge. It is modified according to the properties of its neighbours. This modification may be repeated until all the crack edges of the given image are reviewed to be a part of the border or not. The main steps are described here.

1. Crack edges are partitioned into different patterns according to the types of their two end vertices. The type of a vertex is the number of crack edges that emanate from it. The type of an edge is represented by a pair of numbers consisting of the types of its vertices. For example, in Figure 4.4, the pixel vertices (i, j) and $(i, j+1)$ are the vertices of the crack edge $e_\rightarrow$, and if the types of vertices (i, j) and $(i, j+1)$ are k and l, respectively, the pattern of the edge $e_\rightarrow$ is $(k-1)$.

The type of a vertex u is computed according to its other three crack edges, excluding the one currently being dealt with. Let (a, b, c) be the current confidences of the three crack edges and, without loss of generality, assume that $a \geq b \geq c$. Let q be a constant usually chosen as 0.1, and $m = \max(a, b, c, q)$. Four types of confidences can be computed in association with the vertex u:

$$conf(0) = (m-a)(m-b)(m-c)$$
$$conf(1) = a(m-b)(m-c)$$
$$conf(2) = ab(m-c) \quad (4.8)$$
$$conf(3) = abc$$

Then the type of the vertex u is defined as

$$type(u) = j \quad \text{such that} \quad conf(j) = \max_k conf(k) \quad (4.9)$$

The parameter m adjusts the vertex classification so that it is relative to the local maximum, and the parameter q forces weak vertices to type zero.

Example 4.1 Assuming u and v are two pixel vertices of the crack edge e, the confidences for u and v are given by $(a_u, b_u, c_u) = (0.9, 0.9, 0.01)$ and $(a_v, b_v, c_v) = (0.01, 0.01, 0.01)$, respectively. Check the type of the crack edge e.
Solution: First determine the type of the vertex u. Using Equation 4.8, $conf(2)$ has the largest value. Therefore, $type(u) = 2$, that is, u is a type 2 vertex.

Secondly, determine the type of the vertex v. Similarly, using Equation 4.8, $conf(0)$ shows the largest value. Therefore $type(v) = 0$, that is, v is a type 0 vertex.

Hence, the type of the edge e is 2-0. By symmetry, type 2-0 is considered the same as type 0-2. ∎

2. Every crack edge is assigned with a confidence value as a part of the border, and the edge types are used to modify this confidence. By symmetry, only the following edge types need to be considered:

   0-0: Isolated edge; the edge confidence needs to be decreased.

   0-1: Uncertain; no influence on the edge confidence.

   0-2, 0-3: Dead end; the edge confidence needs to be decreased.

   1-1: Continuation; the edge confidence needs to be increased.

   1-2, 1-3: Continuation; the edge confidence needs to be increased.

   2-2, 2-3, 3-3: Bridge between borders; no influence on the edge confidence.

3. For each crack edge e, an iterative update may be applied to obtain its confidence, denoted as $c(e)$. Superscript (k) is used to denote the k-th iterative update. The $(k + 1)$-th iterative update of edge confidence is based on the edge type and the previous confidence $c^{(k)}(e)$ according to the following choices:

Confidence increases (according to the edge type):
$c^{(k+1)}(e) = \min\{1, c^{(k)}(e) + \delta\}$
Confidence decreases (according to the edge type):
$c^{(k+1)}(e) = \max\{0, c^{(k)}(e) - \delta\}$
No influence (according to the edge type):
$c^{(k+1)}(e) = c^{(k)}(e)$

Here δ denotes a constant chosen in the range from 0.1 to 0.3 [1], which stands for the influence on the edge confidence.

As a summary, the edge relaxation algorithm based on crack edges is given as Algorithm 4.3.

Algorithm 4.3: Edge relaxation based on crack edges
For the given image $f(i, j): 0 \le i \le m-1, 0 \le j \le n-1$
Preparing two thresholds $\tau_1 < \tau_2$ for convergence estimation;
Computing all crack edges for all pixels and initialize edge confidence:
$c^{(0)}(e) = $ normalize $(M(e))$ for every crack edge e;
$k = 0$;
Repeat
{ count=0;
    For each crack edge e:-
    If $(c^{(k)}(e) \ne 0)$ and $c^{(k)}(e) \ne 1$ then
    { count=count + 1;
     Find the edge type according to the confidences of its neighbouring crack edges;
     Update the confidence $c^{(k+1)}(e)$ according to the edge type and $c^{(k)}(e)$;
     If $c^{(k+1)}(e) > \tau_2$ then $c^{(k+1)}(e) = 1$;
     If $c^{(k+1)}(e) < \tau_1$ then $c^{(k+1)}(e) = 0$;
     $k = k + 1$;
    }
} Until (count =0)
End-Algorithm

4.2.3 Border Tracing

As described in section 4.2.1, processing the gradient image with thresholding usually results in images with wider borders. Canny detector, as discussed in Section 4.2.1, is a method of thinning the border. In this section, another method of obtaining borders with one pixel width through border tracing is discussed. The gradient image with thresholding may be presented as a binary image. The first step in the border tracing method is to select an initial border point from the object of a binary image, followed by a search of its 4-neighbouring or 8-neighbouring pixels, and finally output the next border point. To avoid deadlock, a variable is used to record the search direction. Take the 8-neighbourhood shown in Figure 4.7 as an example, integers from 0 to 7 are used to record the different directions of the neighbourhood of the pixel. The search direction is indexed as 0 when the current point is the pixel at (i, j) and the pixel to be searched is $(i, j+1)$. On the other hand, when the next pixel to be searched is located at $(i-1, j+1)$, the search direction is indexed as 1.

Using the image shown in Figure 4.8 as an example, the object point p_0 at the top left corner of the object is selected as the initial border point. In order to describe the process, the variable *dir* is used to record the search direction. The initial search direction is 7, that is, $dir = 7$. The next border point is selected from the 3 × 3-neighbourhood of p_0. The search direction begins with an odd number less than the previous direction 7. Hence, the search starts from direction 5 in an anticlockwise direction. The first object point p_1 found is selected as the new border point, followed by the update $dir = 5$. The next search is in the neighbourhood of p_1. The search begins from direction 3 in an anticlockwise direction to direction 6 where another border point p_2 is obtained. This search process may be repeated until the closed border $p_0 p_1 p_2 \cdots p_9 p_{10} p_0$ is constructed. The tracing algorithm just discussed is summarized in Algorithm 4.4. Note that the border of the *k*-th region is denoted by $p(k, s)$, $s = 0, 1, 2, \ldots, l$, and the output of the algorithm is the sets of pixels that consist of borders of objects.

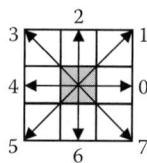

FIGURE 4.7 Directions of search in an 8-neighbourhood.

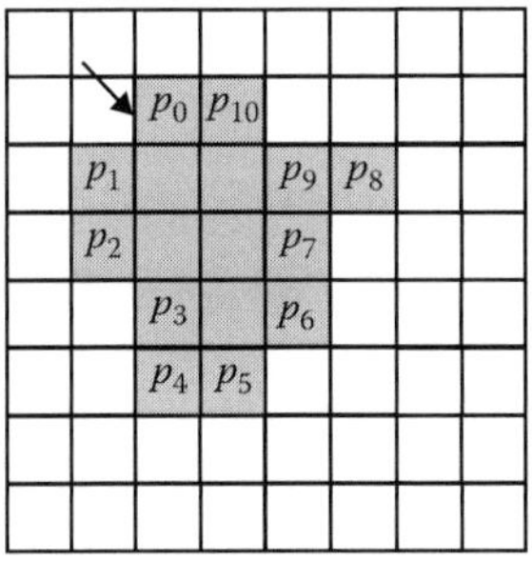

FIGURE 4.8 Border tracing.

Algorithm 4.4: Border tracing detection in 8-neighbourhood
For the given binary image $f(i, j): 0 \leq i \leq m-1; 0 \leq j \leq n-1$;
$k = 0$ denotes the number of the borders;
$q_k = \emptyset$; //the set of the pixel points on the border of the object k;
$si_k = 0$; $sj_k = 0$; // the search beginning point of the k-th region;
do while (true)
{   For $i = si_k$ to $m - 1$ // find a starting border pixel of a new region;
        For $j = sj_k$ to $n - 1$
            if $(f(i, j) = 1)$ and $((i, j) \notin q_s, s = 0, ..., k - 1)$ then
            {$p(k, 0) = (i, j)$; // the first border pixel of the k-th region;
            $si_{k+1} = i$; $sj_{k+1} = j + 1$ // the search of next region will begin from this pixel;
            exit;
            }
            else halt;
        End-For
        End-For;
        Initialize the search direction variable dir = 7;
        $s = 0$;   // the current border pixel;
        Repeat // search border pixels;
        {   If (dir is odd) then dir = (dir + 6) / mod 8
            else dir = (dir + 7) / mod 8; // the beginning direction of search;
            while (dir < 8) and (the corresponding neighbouring pixel is not a border pixel)
            do {dir = (dir + 1) / mod 8;}
            $s = s + 1$;
            $p(k, s)$ = the corresponding neighbouring pixel; $q_k = q_k \cup \{p(k, s)\}$
        } until $p(k, s) = p(k, 1)$ and $p(k, s - 1) = p(k, 0)$;
        $k = k + 1$;

}
End-Algorithm

Note that when $p(k, l) = p(k, 0)$, the border is closed. Figure 4.9 depicts the resulting edge image detected by edge tracing method.

4.2.4 The Hough Transform

Given the shape and size of an edge, the edge point might be obtained more easily by transferring the initial image space to a new space through some kind of transformation. The Hough transform [1,10] is an effective method based on this idea.

It assumes that the image to be processed is a binary image that has been processed using a threshold. Straight lines passing through the point (x, y) can be expressed in form $y = kx + b$, where k denotes the slope, and b the intercept. Different values of k and b govern different lines, that is, any straight line in x–y space is represented by a single point in the k–b parameter space. This single point is relative to the original coordinates (x, y) of any point in the line:

$$b = -kx + y \qquad (4.10)$$

Note that the slope k of a vertical line is infinite, which creates some difficulties in practice. The following polar coordinates transformation may overcome this problem:

$$\rho = x\cos\theta + y\sin\theta \qquad (4.11)$$

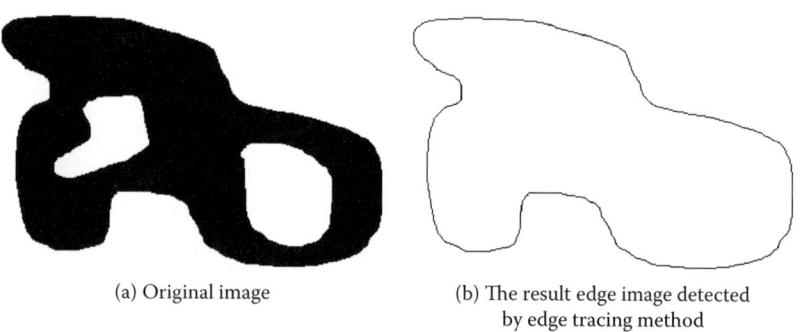

(a) Original image

(b) The result edge image detected by edge tracing method

FIGURE 4.9 Edge detection by the edge-tracing method. (a) Original image, and (b) the resulting edge image detected by edge-tracing method.

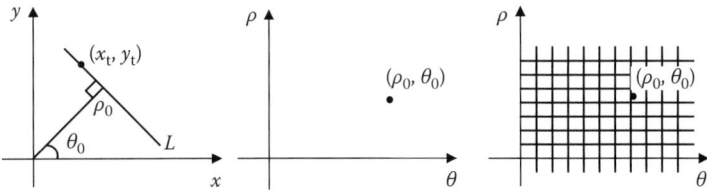

FIGURE 4.10 An illustration of the Hough transform.

By using Equation 4.11, a straight line L in x–y space is transformed to the point (ρ, θ) in the polar coordinates space. As illustrated in Figure 4.10, any point (x_t, y_t) on L is transformed to the same point (ρ_0, θ_0):

$$\rho_0 = x_t \cos\theta_0 + y_t \sin\theta_0 \tag{4.12}$$

If there are n pixel points in the original image that are transformed to the same point (ρ_0, θ_0) in the polar coordinates space, these n pixel points could be a straight line in the original image. The larger the value of n, the more points on the line and the more the possibility that it is the border of a given region of the original image. In the polar coordinate space, an accumulator $A(\rho, \theta)$ is set to count the number of pixel points in the original image that has been transformed to the point (ρ, θ) in polar coordinates space. If $A(\rho, \theta)$ achieves its maximum at a certain point (ρ_1, θ_1), all the pixel points in the original image that have been transformed to (ρ_1, θ_1) make up the line border. The Hough transform is presented in Algorithm 4.5.

Algorithm 4.5: Line detection using the Hough transform
Given the binary image $f(i, j)$: $0 \leq i \leq m-1$, $0 \leq j \leq n-1$
Quantize the parameter space (ρ, θ): $\rho_{min} \leq \rho \leq \rho_{max}$, $0 \leq \theta \leq 180$ where ρ and θ are intergers;
Initialize the accumulator $A(\rho, \theta) = 0$ for all
(ρ, θ): $\rho_{min} \leq \rho \leq \rho_{max}$, $0 \leq \theta \leq 180$.
For $i = 0$ to $m - 1$
For $j = 0$ to $n - 1$
{{

    For $\theta = 0$ to 180
    { $\rho = i\cos\theta + j\sin\theta$
    quantize ρ to the quantisation value ρ';
    $A(\rho',\theta) = A(\rho',\theta) + 1$;
    }

}}

Obtain the maximal value of the accumulator:

$A(\rho^*,\theta^*) = \max\{A(\rho,\theta): p_{min} \leq \rho \leq p_{max}, 0 \leq \theta \leq 180\};$
For $i = 0$ to $m - 1$
For $j = 0$ to $n - 1$
{{
If ($\rho^* = i\cos\theta^* + j\sin\theta^*$) then mark (i, j) as a pixel of the line border.
}}
End-Algorithm

4.3 REGION-BASED SEGMENTATION

Region-based segmentation is different from edge-based segmentation, which extracts the edge of the region before constructing the region. It is a method that directly constructs the region applying a certain homogeneity principle [11,12]. Methods include the region-growing method, image-merging method, and region split-and-merge method.

4.3.1 The Region-Growing Method

Region growing picks up one pixel of the image as a seed to start with. The initial region contains only the seed, which is then compared with its neighbouring pixels according to a certain homogeneity principle before adding any analogical pixels to the initial region. The same process is repeated until the region stops growing.

The homogeneity principle may be based on grey level, colour, texture, shape, model, etc. For example, the magnitude of crack edges may be used as a suitable metric. Suppose v is the current point in the region, and u is a pixel point in its 4-neighbourhood to be compared; a homogeneity criterion for the given image function f can be defined as

$$s = \begin{cases} 1, & \text{if } |f(u) - f(v)| < \tau \\ 0, & \text{if } |f(u) - f(v)| \geq \tau \end{cases} \quad (4.13)$$

where τ is a preassigned threshold. The case $s = 1$ means u and v are similar and may be put in the same region; otherwise, u and v belong to different regions. The following algorithm shows the computational steps of the region-growing method.

Algorithm 4.6: The region-growing method
For the given image $f(i,j): 0 \leq i \leq m-1, 0 \leq j \leq n-1$
Preassign a seed pixel v_0 and a threshold τ;
Initialize the current region $R := \{v_0\}$; The set of candidate seeds $C := \{v_0\}$;

While $(C \neq \varnothing)$ do
{
    select a seed pixel v from C;
    $C := C \setminus \{v\}$; // delete the seed pixel v from the candidate seed set;
    $N(v) :=$ the set containing 4-neighbouring pixels of v;
    For each $u \in N(v)$: if $(u \notin R)$ and $(s = 1)$ // From Equation 4.13
    {
        $R := R \cup \{u\}$;
        $C := C \cup \{u\}$ // u is added to the region and is a candidate seed
    }
}
End-Algorithm

4.3.2 The Region-Merging Method

Region merging begins with partitioning the original image into small regions, followed by the combination of similar adjacent regions into a bigger region according to a given homogeneity principle. The process of merging is repeated until each region is the largest and does not grow any more in accordance with the homogeneity principle. Furthermore, either the mean grey-level value or some other statistical characteristic of each adjacent region is to be used to establish comparability between regions. The difference between this method and the method of region growing lies in the use of comparability. Statistical measures such as variance, grey-level histogram, etc., can be used in establishing comparability.

The following algorithm shows the computational steps involved in the method of region merging, in which the homogeneity principle for regions is based on the difference between the mean grey-levels of two regions.

Algorithm 4.7: Region merging
Given the image $f(i, j) : 0 \leq i \leq m-1, 0 \leq j \leq n-1$
$k = 0$; // current number of regions;
initialize all pixels as unmarked;
For $i = 0$ to $m - 1$
For $j = 0$ to $n - 1$
// Partition the original image into regions of constant grey-level.
{{ if (pixel (i, j) is unmarked) then
    mark (i, j);
    for each 8-neighbouring pixel u:
        case $(f(u) = f(i, j))$ and (u is marked):

(i, j) belongs to the region which u belongs to;
case (f(u) = f(i, j)) and (u is unmarked):
    (i, j) and u belong to the same region R_k;
    k: = k + 1; mark u;
case (f(u) ≠ f(i, j)):
    (i, j) belongs to the region R_k; k: = k + 1;
end-for;
}}
n_merge = 0; // tag of merges;
Repeat
{ For all region R_s, compute r_s = the mean grey level of R_s;
    For each region R_s, compare R_s with its neighbour region R_t:
        if $|r_s - r_t| < T$ then
            combine R_s, R_t to form a new region;
            n_merge = n_merge + 1;
        endif
} until (n_merge: = 0)
End-Algorithm

4.3.3 The Region Split-and-Merge Method

Region split-and-merge is also used in image segmentation. First, an image is treated as a bigger region, which is divided into smaller regions according to homogeneity principles such as variance, grey-scale histogram etc. Second, similar adjacent smaller regions are merged by checking the comparability of these regions. Repeat the merging process until the region cannot grow any more. Particular attention should be paid to different homogeneity principles of split and merge.

4.4 FURTHER READING

Several basic image segmentation methods are introduced in this chapter. Thresholding methods rely on the intensities of the pixels only and neglect the variation along spatial positions. These methods are particularly suitable for images with distinguishable background and foreground. They do not work when the edges are blurry. In other words, threshold techniques are effective only if all pixels that belong to the objects have brightness levels within a certain range that can be distinguished from those of the background. Edge-based segmentation depends on edge detection and edge acquisition. However, it is difficult to detect a continuous closed curve that encircles a region. Region-based segmentation depends on region comparability and the choice of a homogeneity criterion. Larger regions are segmented into smaller ones if the criterion is too strong, whereas different regions are combined to form a larger region if the criterion is too weak [13].

The results of image segmentation are mainly used in image recognition and image understanding. It is not enough to process only low-level data such as intensities or spatial locations. Prior and professional knowledge of images (such as medical images) is helpful in obtaining a precise segmentation. Most of the algorithms for image segmentation depend on searching methods. There are many search algorithms that use the graph-based approach [14,15], multiscale approach [16,17], neural network approach [18], dynamic programming approach [19], genetic algorithm approach [20], etc. Apart from searching methods, readers may wish to read more about contemporary methods in image segmentation that use the clustering method [21] and matching method [22].

4.5　EXERCISES

Q.1　The histogram of a 4-bit grey-scale image having a size 16 × 16 is described as the following array. Select a threshold to segment the foreground from the background.

(0 5 5 10 16 10 5 10 20 35 55 35 25 15 10 0)

Q.2　Detect the border of the picture given in Figure 4.11 using Algorithm 4.4 ("Border tracing detection in 8-neighbourhood").

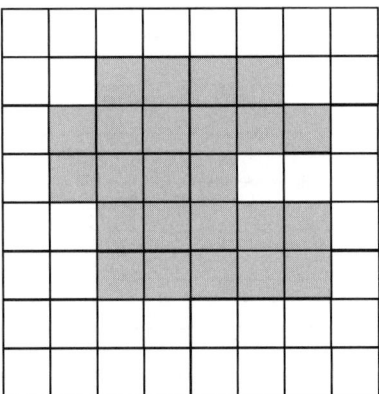

FIGURE 4.11　Q.2.

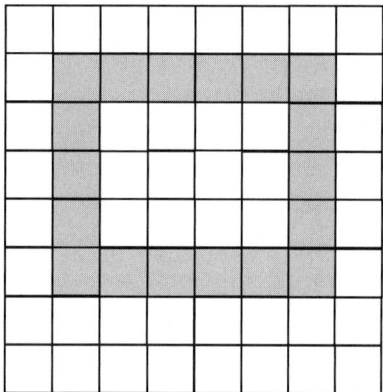

FIGURE 4.12 Q.4.

Q.3 Using different grey-scale images, compare the Canny edge detection method with other edge image thresholding detection methods with which edge images may be obtained by the Roberts operator, Prewitt operator, Sobel operator, or Laplacian operator, as defined in Chapter 3, Section 3.4.3.

Q.4 Find the Hough transform of the shaded region given in Figure 4.12.

Q.5 Apply the region-growing method described in Algorithm 4.6 to find the objects in Figure 4.13.

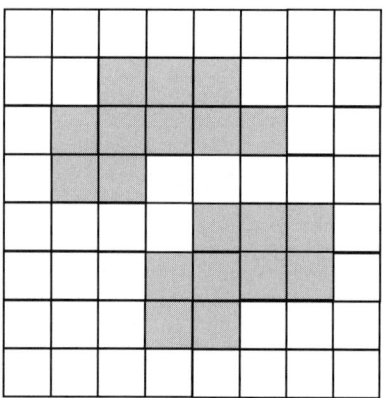

FIGURE 4.13 Q.5.

4.6 REFERENCES

1. M. Sonka, V. Hlavac, and R. Boyle. *Image Processing: Analysis and Machine Vision*, 2nd Edition, Thomson Learning and PPTPH, Monterey, CA, 1998.
2. K. R. Castleman. *Digital Image Processing*, Prentice Hall, Englewood Cliffs, NJ, 1998.
3. S. D. Yanowitz and A. M. Bruckstein. A new method for image segmentation. Computer Vision. *Graphics and Image Processing*, 46: 82–95, 1989.
4. T. W. Rilder and S. Calvard. Picture thresholding using an iterative selection method. *IEEE Transactions on Systems, Man and Cybernetics*, 8(8): 630–632, 1978.
5. J. F. Canny. A computational approach to edge detection. *IEEE Transaction on Pattern Analysis and Machine Intelligence*, 8(6): 679–698, 1986.
6. J. Yunde. *Machine Vision*, Science Publication, Beijing, China, 2000 (in Chinese).
7. B. Morse. Segmentation. Available at http://homepages.inf.ed.ac.uk/rbf/CVonline/LOCAL_COPIES/MORSE/edgeseg.pdf. In CVonline: On-Line Compendium of Computer Vision (Online). R. Fisher (Ed.). Available: http://homepages.inf.ed.ac.uk/rbf/CVonline/.
8. H. D. Ballard and M. C. Brown. *Computer Vision*, Prentice-Hall, Englewood Cliffs, NJ, 1982.
9. J. M. Prager. Extracting and labelling boundary segments in natural scenes. *IEEE Transactions on Pattern Analysis and Machine Intelligence*, 2(1): 16–27, 1980.
10. R. O. Duda and P. E. Hart. *Pattern Classification and Scene Analysis*, John Wiley, New York, 1973.
11. R. M. Haralick and L. G. Shapiro. Image segmentation techniques. *Computer Vision, Graphics, and Image Processing*, 29: 100–132, 1985.
12. Y. L. Chang and X. Li. Fast image region growing. *Image and Vision Computing*, 13: 559–571, 1995.
13. T. Asano, D. Z. Chen, N. Katoh, and T. Tokyama. Polynomial-time solution to image segmentation. *Proceedings of the 7th Annual. SIAM-ACM Conference on Discrete Algorithm*, 104–113, 1996.
14. N. J. Nilsson. *Principles of Artificial Intelligence*, Springer Verlag, Berlin, 1982.
15. J. Shi and J. Malik. Normalized cuts and image segmentation. *IEEE Transactions on Pattern Analysis and Machine Intelligence*, 22(8): 888–905, 2000.
16. L. M. Lifshitz and S.M. Pizer. A multiresolution hierarchical approach to image segmentation based on intensity extrema. *IEEE Transaction on Pattern Analysis and Machine Intelligence*, 12(6): 529–540, 1990.
17. M. Tabb and N. Ahuja. Unsupervised multiscale image segmentation by integrated edge and region detection. *IEEE Transactions on Image Processing*, 6(5): 642–655, 1997.
18. M. Egmont-Petersen, D. de Ridder, and H. Handels. Image processing with neural networks—a review. *Pattern Recognition*, 35: 2279–2301, 2002.
19. W. A. Barrett and E. N. Mortensen. Fast, accurate, and reproducible live-wire boundary extraction. *Visualization in Biomedical Computing*, pp. 183–192, Springer Verlag, Berlin, 1996.

20. G. S. Seetharaman, A. Narasimhan, A. Sathe, and L. Storc. Image segmentation with genetic algorithms: a formulation and implementation. *Stochastic and Neural Methods in Signal Processing, Image Processing, and Computer Vision*, Proceedings of SPIE, Vol. 1569: 269–273, 1991.
21. A. K. Jain and P. J. Flynn. Image segmentation using clustering. *Advances in Image Understanding: A Festschrift for Azriel Rosenfel*, pp. 65–83, 1996.
22. D. M. Gavrila and L. S. Davis. Towards 3D model-based tracking and recognition of human movement: A multi-view approach. *Proceedings of The International Workshop on Face and Gesture Recognition*, Zurich, 1995.

4.7 PARTIAL CODE EXAMPLES

Project 4-1: Optimal Thresholding Segmentation

(These codes can be found in CD: Project4-1\source code\project4-1 View.cpp)

```
#include "stdafx.h"
#include "project4_1.h"
#include "project4_1Doc.h"
#include "project4_1View.h"
#ifdef _DEBUG
#define new DEBUG_NEW
#undef THIS_FILE
static char THIS_FILE[] = __FILE__;
#endif
/*****************************************************
 *********
 * Function name:
 * OnOptimalthresholding()
 *
 * Parameter:
 * HDIB hDIB - the handle of the image
 *
 * Return Value:
 * None
 *
 * Description:
 * Optimal thresholding
 *
 *****************************************************
 *******/
void CProject4_1View::OnOptimalthresholding()
{
```

Image Segmentation ■ 159

```
      int i,j,T0=0,T1=0,u1,u2,tmp1,tmp2;
unsigned char *lpSrc;
      CProject4_1Doc* pDoc = GetDocument();
      ASSERT_VALID(pDoc);
      if(pDoc->m_hDIB == NULL)
            return ;
      LPSTR lpDIB = (LPSTR) ::GlobalLock((HGLOBAL)
pDoc->m_hDIB);
      LPSTR lpDIBBits=::FindDIBBits(lpDIB);
      int cxDIB = (int) ::DIBWidth(lpDIB); // Size of
DIB - x
      int cyDIB = (int) ::DIBHeight(lpDIB); // Size of
DIB - y
      long lLineBytes = WIDTHBYTES(cxDIB * 8); // count
the number of byte of the image per line
      // per line
      for(i = 0; i < cyDIB; i++)
      {
            // per column
            for(j = 0; j < cxDIB; j++)
            {
                  // the pointer pointing to the i-th
line and j-th picture element
                  lpSrc = (unsigned char*)lpDIBBits +
lLineBytes * (cyDIB - 1 - i) + j;

                  // computing the value of gradation
                  T0=T0+*lpSrc;
            }
      }
      T0=T0/(cyDIB*cxDIB);
      while(1)
      {
            u1=0;u2=0;
            tmp1=0;tmp2=0;
            // per line
            for(i = 0; i < cyDIB; i++)
            {
                  // per column
                  for(j = 0; j < cxDIB; j++)
                  {
```

```
                        // the pointer pointing to the i-th line and j-th picture element
                        lpSrc = (unsigned char*)lpDIBBits + lLineBytes * (cyDIB - 1 - i) + j;
                        if (*lpSrc>=T0)
                        {
                                u1=u1+*lpSrc;
                                tmp1=tmp1+1;
                        }
                        else
                        {
                                u2=u2+*lpSrc;
                                tmp2=tmp2+1;
                        }
                }
        }
        u1=u1/tmp1;
        u2=u2/tmp2;
        T1=(u1+u2)/2;
        if (T0==T1)
                break;
        T0=T1;
}
// per line
for(i = 0; i < cyDIB; i++)
{
        // per column
        for(j = 0; j < cxDIB; j++)
        {
                // the pointer pointing to the i-th line and j-th picture element
                lpSrc = (unsigned char*)lpDIBBits + lLineBytes * (cyDIB - 1 - i) + j;

                // computing the value of gradation
                if(*lpSrc<=T0) *lpSrc=BYTE(0);
                else *lpSrc = BYTE(255);
        }
}
    ::GlobalUnlock((HGLOBAL) pDoc->m_hDIB);
 Invalidate(TRUE);
}
```

Project 4-2: The Border-Tracing Method

(These codes can be found in CD: Project4-2\source code\project4-2 View.cpp)

```
#include "stdafx.h"
#include "project4_2.h"
#include "GlobalApi.h"
#include "project4_2Doc.h"
#include "project4_2View.h"
#include "math.h"
#ifdef _DEBUG
#define new DEBUG_NEW
#undef THIS_FILE
static char THIS_FILE[] = __FILE__;
#endif
/*************************************************
*********
* Function name:
* OnBorderTracing()
*
* Parameter:
* None
*
* Return Value:
* None
*
* Description:
* Border tracing
*
*************************************************
*******/
void CProject4_2View::OnBorderTracing()
{
        //change the style of cursor
        BeginWaitCursor();
//  unsigned char *lpSrc;
        CProject4_2Doc* pDoc = GetDocument();
        ASSERT_VALID(pDoc);
        if(pDoc->m_hDIB == NULL)
                return ;
        LPSTR lpDIB = (LPSTR) ::GlobalLock((HGLOBAL) pDoc->m_hDIB);
```

```
        LPSTR lpDIBBits=::FindDIBBits (lpDIB);
        int cxDIB = (int) ::DIBWidth(lpDIB); // Size of
DIB - x
        int cyDIB = (int) ::DIBHeight(lpDIB); // Size of
DIB - y
        long lLineBytes = WIDTHBYTES(cxDIB * 8); // count
the number of byte of the image per line
        // the pointer pointing to the source image
        LPSTR lpSrc;

        // the pointer pointing to the buffer image
        LPSTR lpDst;

        // the pointer pointing to the buffer DIB image
        LPSTR lpNewDIBBits;
        HLOCAL hNewDIBBits;

        // cycle variants
        long i;
        long j;
        int lWidth = cxDIB ;
        int lHeight= cyDIB ;
        // intensity of a pixel
        unsigned char pixel;
        // the tag used for marking the start point
        bool bFindStartPoint;
        //the tag used for marking a border point
        bool bFindPoint;
        //the start border point and the current border
point
        Point StartPoint,CurrentPoint;
        // eight directions and the initial scanning
direction
        int Direction[8][2]={{-1,1},{0,1},{1,1},{1,0},
{1,-1},{0,-1},{-1,-1},{-1,0}};
        int BeginDirect;
        // allocate memory for the new image
        hNewDIBBits = LocalAlloc(LHND, lLineBytes *
lHeight);

        // lock memory
        lpNewDIBBits = (char * )LocalLock(hNewDIBBits);
```

```
      // initialise the allocated memory with the constant 255
      lpDst = (char *)lpNewDIBBits;
      memset(lpDst, (BYTE)255, lLineBytes * lHeight);
      //first find the border point in the top left
      bFindStartPoint = false;
      for (j = 0;j < lHeight && !bFindStartPoint;j++)
      {
            for(i = 0;i < lWidth && !bFindStartPoint;i++)
            {
      // the pointer pointing to the i-th line and j-th column picture pixel
      // from the bottom
                  lpSrc = (char *)lpDIBBits + lLineBytes * j + i;

                  // get the intensity of the current pointer and convert it to unsigned char
                  pixel = (unsigned char)*lpSrc;

                  if(pixel == 0)
                  {
                        bFindStartPoint = true;
                        StartPoint.Height = j;
                        StartPoint.Width = i;
      // the pointer pointing to the i-th line and j-th column pixel of the // destination image from the bottom
                        lpDst = (char *)lpNewDIBBits + lLineBytes * j + i;
                        *lpDst = (unsigned char)0;
                  }
            }
      }
      // initial scanning direction
      BeginDirect = 0;
      //trace the border
      bFindStartPoint = false;
      // begin to scan from the initial scanning direction in the start point
      CurrentPoint.Height = StartPoint.Height;
      CurrentPoint.Width = StartPoint.Width;
      while(!bFindStartPoint)
```

```cpp
        {
            bFindPoint = false;
            while(!bFindPoint)
            {
                //check a pixel along the scanning direction
                lpSrc = (char *)lpDIBBits + lLineBytes * (CurrentPoint.Height + Direction[BeginDirect][1])
                    + (CurrentPoint.Width + Direction[BeginDirect][0]);
                pixel = (unsigned char)*lpSrc;
                if(pixel == 0)
                {
                    bFindPoint = true;
                    CurrentPoint.Height = CurrentPoint.Height + Direction[BeginDirect][1];
                    CurrentPoint.Width = CurrentPoint.Width + Direction[BeginDirect][0];
                    if(CurrentPoint.Height == StartPoint.Height && CurrentPoint.Width == StartPoint.Width)
                    {
                        bFindStartPoint = true;
                    }
                    lpDst = (char *)lpNewDIBBits + lLineBytes * CurrentPoint.Height + CurrentPoint.Width;
                    *lpDst = (unsigned char)0;
                    // rotate the scanning direction two steps along anti-clock direction
                    BeginDirect--;
                    if(BeginDirect == -1)
                        BeginDirect = 7;
                    BeginDirect--;
                    if(BeginDirect == -1)
                        BeginDirect = 7;
                }
                else
                {
                    // rotate the scanning direction one step along the clock direction
                    BeginDirect++;
                    if(BeginDirect == 8)
```

```
                                BeginDirect = 0;
            }
        }
}
// copy the new image
memcpy(lpDIBBits, lpNewDIBBits, lWidth * lHeight);
// free memory
LocalUnlock(hNewDIBBits);
LocalFree(hNewDIBBits);
// restore the style of the cursor
EndWaitCursor();

// set modified flag
pDoc->SetModifiedFlag(TRUE);

// update all views
pDoc->UpdateAllViews(NULL);
}
```

CHAPTER 5

Mathematical Morphology

Mathematical morphology was initially developed to analyse the shape and structure of objects [1] in binary images. In particular, it is a useful tool for extracting important components of a binary image, leading to easier image representation and description. Its concepts and mathematical operations, which come through from the set theory, are quite different from the treatises in Chapters 3 and 4. In these chapters, the methods of image processing focus on the intensity functions of the images. Concepts and processing techniques used in mathematical morphology in this chapter are aimed at the use of set operations. Note that these concepts can be extended to handle image preprocessing and image segmentation as described in Chapters 3 and 4. In summary, techniques employed in Chapters 3 and 4 are based on point-spread function and linear transformations such as convolution [2], whereas the basic ingredient in mathematical morphology is set theory.

In this chapter, basic concepts and operations of mathematical morphology for binary and grey-scale images are given. Examples of handling binary images and elementary operations used in mathematical morphology are included. Rigorous mathematics has been avoided in this chapter. However important algorithms are presented for binary images with extensions to grey-scale images. Details of set operations and their equivalent computer implementations are also presented. Images resulting from the applications of mathematical morphology are included in this chapter.

5.1 SOME BASIC CONCEPTS OF SET THEORY

The concepts and operations involved in mathematical morphology come from the set theory. In this section, some basic concepts of the set theory [3] are briefly overviewed.

5.1.1 Sets and Elements

A set A is a collection of elements having the same property. An element a of A is denoted as $a \in A$. If an element x does not belong to the set A, it is denoted as $x \notin A$. An empty set $\varnothing$ is a set that contains null element.

Example 5.1 Let Z be the set of all integers, then $5 \in Z$, but $2.5 \notin Z$.

Example 5.2 The object of a binary image can be considered as a set consisting of the coordinates of pixels with the intensity 0 (i.e., the background is white).

5.1.2 Relationships between Two Sets

A set A is equal to another set B if A and B consist of exactly the same elements. This relationship is written as $A = B$; otherwise, $A \neq B$.

A is known as a subset of B if for all $a \in A$, $a \in B$. A is said to be contained in B and is denoted as $A \subseteq B$ or B contains A and is denoted as $B \supseteq A$. Suppose $A \subseteq B$ and $A \neq B$, A is called a proper subset of B and is denoted as $A \subset B$ or $B \supset A$.

5.1.3 Operations Involving Sets

Given two sets A and B, which are contained in the universal set S:

(a) $A \cup B$, the union of A and B, is a new set defined as follows:

$$A \cup B = \{x \mid x \in A \quad \text{or} \quad x \in B\} \tag{5.1}$$

(b) $A \cap B$, the intersection of A and B, is a new set given by

$$A \cap B = \{x \mid x \in A \quad \text{and} \quad x \in B\} \tag{5.2}$$

(c) $A - B$, the difference of A and B, is defined as

$$A - B = \{x \mid x \in A \quad \text{and} \quad x \notin B\} \tag{5.3}$$

(d) A^c, the complement of A, is defined as

$$A^c = \{x \mid x \in S \quad \text{and} \quad x \notin A\}$$
$$= S - A \tag{5.4}$$

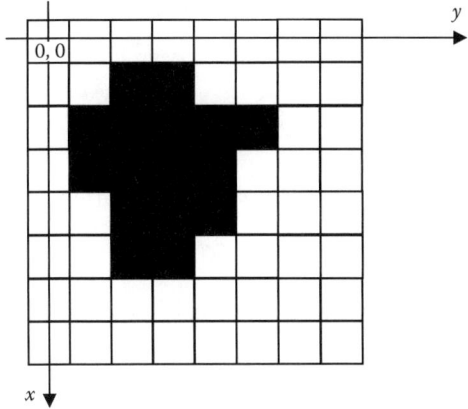

FIGURE 5.1 A binary image as a set.

5.2 MORPHOLOGY FOR BINARY IMAGES

Mathematical morphology was initially used to process binary images, and results were promising. In mathematical morphology, a binary image is treated as a set consisting of the ordered pairs of coordinates of pixel points in that image.

In Figure 5.1, the top left pixel point is the origin, and its coordinates are given as (0,0). The coordinates of the pixel points in the first row from left to right are (0,0), (0,1), ..., (0,7). Similarly, the coordinates of the bottom right pixel point are (7,7). In essence, the given image in Figure 5.1 is considered as a set of ordered pairs of coordinates denoted as $S = \{(i,j) | 0 \leq i \leq 7, 0 \leq j \leq 7\}$. The object in the image consists of black points whose corresponding coordinates are

$$(1,2),(1,3);$$

$$(2,1),(2,2),(2,3),(2,4),(2,5);$$

$$(3,1),(3,2),(3,3),(3,4);$$

$$(4,2),(4,3),(4,4);$$

$$(5,2),(5,3)$$

The set of coordinates corresponding to the object is

$$A = \{(1,2),(1,3),(2,1),(2,2),(2,3),(2,4),(2,5),$$
$$(3,1),(3,2),(3,3),(3,4),(4,2),(4,3),(4,4),(5,2),(5,3)\}$$

Suppose the whole image is considered as a universal set, then the background is the complement of A:

$$A^c = S - A$$

Based on the addition of coordinates, the translation A_h of the set A by the point $h \in s$ is defined as

$$A_h = A + h = \{x + h \in S \mid x \in A\} \tag{5.5}$$

Example 5.3 Figure 5.2 shows the universal set $S = \{(i, j) \mid 0 \le i \le 7, 0 \le j \le 7\}$ and an object $C = \{(1,2),(1,3)\}$. The translation of C by $h = (3,2) \in S$ can be calculated as

$$C_h = \{(1,2)+(3,2),(1,3)+(3,2)\} = \{(4,4),(4,5)\} \quad \blacksquare$$

Another important concept is the *structuring element*. A structuring element E is a set consisting of a local origin o, known as the *representative point*, and its neighbouring points. Figure 5.3 shows some typical structuring elements [2] given by

(1) $E_1 = \{(0,-1),(0,0),(0,1)\}$

(2) $E_2 = \{(0,-1),(0,1)\}$

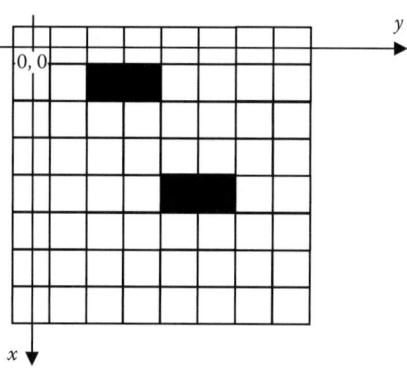

FIGURE 5.2 A set and its translation.

(3) $E_3 = \{(-1,0),(0,0),(1,0)\}$

(4) $E_4 = \{(-1,0),(0,-1,),(0,0),(0,1),(1,0)\}$

In these structuring elements, the representative point is (0,0).

5.2.1 Binary Morphological Operation

The two basic mathematical morphology operations for binary images are *dilation* and *erosion*, from which other complex operations can be defined.

5.2.1.1 Dilation Operation

The dilation of two sets A and B denoted by $A \oplus B$ is defined as

$$A \oplus B = \bigcup_{b \in B} A_b \qquad (5.6)$$

It is easy to prove that the dilation operation is commutative and associative, that is,

$$\begin{aligned} A \oplus B &= B \oplus A \\ A \oplus (B \oplus C) &= (A \oplus B) \oplus C \end{aligned} \qquad (5.7)$$

The dilation operation is often used to process an image with a structuring element. Take Equation 5.6 as an example; A is an image and B is a structuring element. The purpose of performing dilation is to enlarge a given object. Through this process, some unfilled parts within objects may be filled in.

Example 5.4 Figure 5.4 shows an object defined by the set

$$A = \{(1,2),(1,3),(2,1),(2,2),(2,3),(3,1),(3,2),(3,3)\}$$

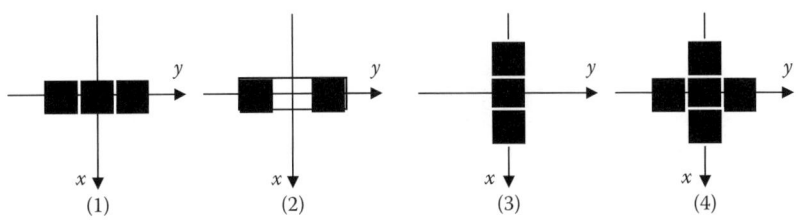

FIGURE 5.3 Some typical structuring elements.

The structuring element $E_2 = \{(0,-1),(0,1)\}$ shown in Figure 5.3 is adopted here. The dilation of the set A and E_2 can be computed as follows:

$$A \oplus E_2 = \bigcup_{b \in B} A_b = A_{(0,-1)} \cup A_{(0,1)}$$

$$= \{(1,2)+(0,-1),(1,3)+(0,-1),(2,1)+(0,-1),(2,2)+(0,-1),(2,3)+(0,-1),$$

$$(3,1)+(0,-1),(3,2)+(0,-1),(3,3)+(0,-1)\} \cup \{(1,2)+(0,1),(1,3)+(0,1),$$

$$(2,1)+(0,1),(2,2)+(0,1),(2,3)+(0,1),(3,1)+(0,1),(3,2)+(0,1),(3,3)+(0,1)\}$$

$$= \{(1,1),(1,2),(1,3),(1,4),(2,0),(2,1),(2,2),(2,3),(2,4),(3,0),(3,1),(3,2),(3,3),(3,4)\}$$

The dilated image is the union of black pixels and grey pixels, as shown in Figure 5.4. The result shows an expansion of A to the left [translation by $(0,-1)$] and to the right [translation by $(0, 1)$]. ■

Let $f(i,j)$, $0 \le i, j \le n-1$, be a binary image with white background such that the object set is defined as $A = \{(i,j) | f(i,j) = 0\}$. Let $e(s,t)$, $-m \le s, t \le m$, be a structure element matrix that defines the structure element set $E = \{(s,t) | e(s,t) = 1, -m \le s, t \le m\}$. The dilation operation can be implemented by using Algorithm 5.1, in which the resulting image is denoted by the image function $g(i,j)$, $0 \le i, j \le n-1$.

Algorithm 5.1: Dilation algorithm
For the given binary image $f(i,j), 0 \le i, j \le n-1$ with
the given structure element array $e(s,t), 0 \le s, t \le m-1$:-
For $i = 0$ to $n = 1$
    For $j = 0$ to $n = 1$ do
        $g(i, j) = 1$;
        For $s = -m$ to m

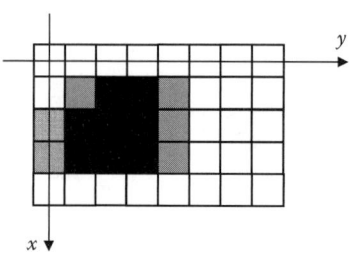

FIGURE 5.4 A dilation example $A \oplus E_2$.

For $t = -m$ to m do
    If $((e(s, t) == 1)$ and $(f(i + s, j + t) == 0))$ then
        $g(i, j) = 0$;
        exit;
    End-If
End-For
If $(g(i, j) == 0)$ exit;
End-For
End-For
End-For

Output of the resulting image: $g(i,j), 0 \leq i, j \leq n-1$
End-Algorithm

5.2.1.2 Erosion Operation

The erosion of two sets A and B is denoted by $A \ominus B$, and is defined as

$$A \ominus B = \bigcap_{b \in B} A_{-b} \qquad (5.8)$$

The effect of erosion is shrinking of an object, and the amount of shrinkage depends on the structuring element.

Example 5.5 Use the object set A as defined in Example 5.3, and assume the structuring element to be $E_2 = \{(0,-1)(0,1)\}$. The result of erosion of A and E_2 is the following set:

$$A \ominus E_2 = \bigcap_{b \in B} A_{-b} = A_{(0,1)} \cap A_{(0,-1)}$$

$$= \{(1,2)+(0,1),(1,3)+(0,1),(2,1)+(0,1),(2,2)+(0,1),(2,3)+(0,1),(3,1)+(0,1),(3,2)$$

$$+(0,1),(3,3)+(0,1)\} \cap \{(1,2)+(0,-1),(1,3)+(0,-1),(2,1)+(0,-1),(2,2)$$

$$+(0,-1),(2,3)+(0,-1),(3,1)+(0,-1),(3,2)+(0,-1),(3,3)+(0,-1)\}$$

$$= \{(1,3),(1,4),(2,2),(2,3),(2,4),(3,2),(3,3),(3,4)\}$$

$$\cap \{(1,1),(1,2),(2,0),(2,1),(2,2),(3,0),(3,1),(3,2)\}$$

$$= \{(2,2),(3,2)\}$$

Figure 5.5 depicts the erosion of A and E_2. The object consisting of only grey pixel points is the result of $A \ominus E_2$. ∎

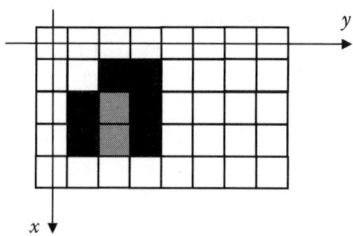

FIGURE 5.5 Example of an erosion operation.

The erosion operation can be implemented by the following algorithm, in which the definitions of $f(i, j)$, $0 \le i, j \le n-1$ and $e(s, t)$, $-m \le s, t \le m$ are the same as in Algorithm 5.1, and $g(i, j)$, $0 \le i, j \le n-1$ is the resulting image of the erosion operation. Figure 5.6 depicts the results of dilation and erosion operations of a binary image.

Algorithm 5.2: Erosion algorithm
For the given binary image $f(i,j), 0 \le i, j \le n-1$ with
the given structure element array $e(s,t), 0 \le s, t \le m-1$:-
For $i = 0$ to $n-1$
    For $j = 0$ to $n-1$ do
        $g(i,j) = 0$;
        For $s = -m$ to m
            For $t = -m$ to m do
                If $((e(s,t)==1)$ and $f(i+s, j+t)==1)$ then
                    $g(i,j) = 1$;
                    exit;
                End-If
            End-For
            If $(g(i,j) == 1)$ exit;
        End-For
    End-For
End-For
Output of the resulting image: $g(i,j), 0 \le i, j \le n-1$
End-Algorithm

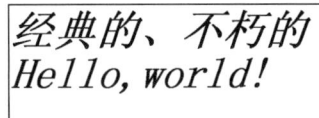

(a) Original image

(b) Dilation result of (a)   (c) Erosion result of (a)

FIGURE 5.6 The results of binary dilation and erosion operations: (a) the original image, (b) dilation result of (a), and (c) erosion result of (a).

5.2.1.3 Opening and Closing Operations

Opening and closing operations are based on dilation and erosion. The opening of a binary image A by the structuring element E is denoted by $A \circ E$ and is defined as

$$A \circ E = (A \ominus E) \oplus E \qquad (5.9)$$

The closing of A by the structuring element E is denoted by $A \bullet E$ and is defined as

$$A \bullet E = (A \oplus E) \ominus E \qquad (5.10)$$

It should be noted that dilation is not an inverse transformation of erosion and vice versa. $A \circ E$ is not the same as $A \bullet E$. Both these operations are often used to smooth the contours of objects. In general, the opening operation weakens the narrow isthmuses and eliminates thin protrusions in images, whereas the closing operation tends to fuse narrow breaks and fill gaps in contours [4]. Figure 5.7 depicts the results of the opening and the closing operations of a binary image.

5.2.1.4 Hit-or-Miss Transformation

The operators defined in the previous sections are used to handle objects with single structuring elements. The hit-or-miss transformation uses two structuring elements simultaneously, one for the object of the given image and the other for the background. Based on this idea, a composite

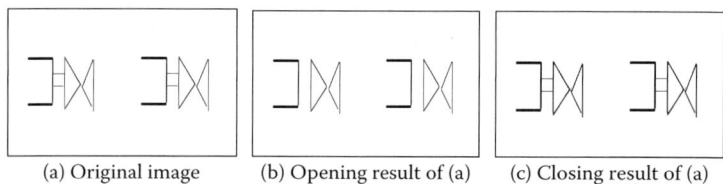

(a) Original image   (b) Opening result of (a)   (c) Closing result of (a)

FIGURE 5.7 The results of binary opening and closing operations: (a) the original image, (b) opening result of (a), and (c) closing result of (a).

structuring element E is required, and it may be defined as a pair of *disjoint structuring elements* [2] such as the following one:

$$E = (E_1, E_2) \tag{5.11}$$

The hit-or-miss transformation of an image A with the composite structuring element E is defined as

$$A \otimes E = (A \ominus E_1) \cap (A^c \ominus E_2) = (\bigcap_{e \in B_1} A_{-e}) \cap (\bigcap_{e \in B_2} A^c_{-e}) \tag{5.12}$$

Equation 5.12 is equivalent to

$$A \otimes E = (A \ominus E_1) - (A \oplus E_2) \tag{5.13}$$

5.2.2 Applications of Binary Morphological Operations

Binary morphology can be used to extract the borders of an object in binary images. Thinning, thickening, and skeleton methods described in following sections are commonly used. Code implementations can be found in Section 5.7 at the end of this chapter.

5.2.2.1 Thinning and Thickening

Thinning operation is often used to make lines in images having more than one-pixel width thinner, whereas thickening operation is used to broaden the lines that may connect broken borders.

 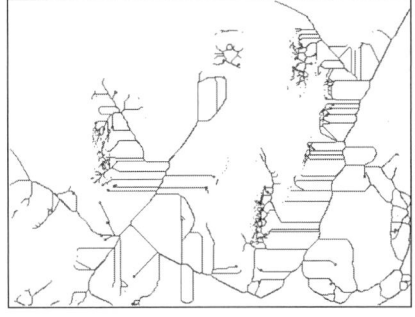

(a) Original image   (b) Thinning result of (a)

FIGURE 5.8 The result of a binary thinning operation: (a) the original image, and (b) the result after thinning.

The thinning and thickening operations of an image A with the composite structuring element E can be described by the following set of operations:

$$\text{Thinning: } \square = A - (A \otimes E) \tag{5.14}$$

$$\text{Thickening: } \boxtimes = A \cup (A \otimes E) \tag{5.15}$$

For a given binary image, a thinning or thickening operation may be repeated several times in order to obtain a good result. The typical result of a binary thinning operation is shown in Figure 5.8.

5.2.2.2 Skeleton Method

A skeleton is known as the medial axis of an object, and is a one-pixel thick line through the middle of the object, preserving the topology of the object [5, 6, 7]. A skeleton can represent the shape of an object, and is commonly used as a feature of objects in image analysis and image recognition.

Denote $A \ominus kE$ as the result of an image A eroded by the structuring element E, k times, that is,

$$A \ominus kE = (A \ominus (k-1)E) \ominus E, k = 2, 3, ..., K \tag{5.16}$$

where K satisfies the condition.

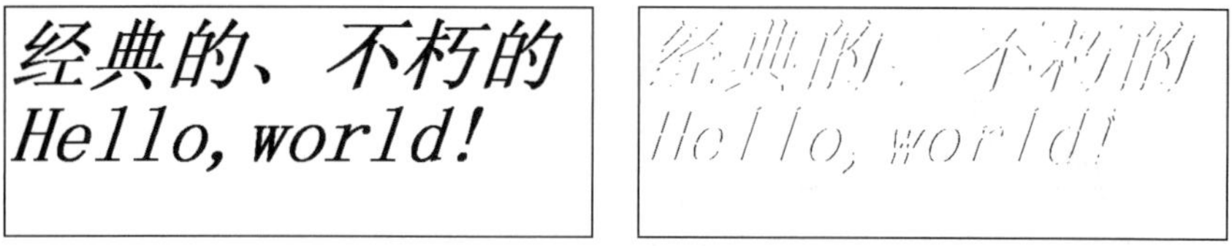

(a) Original image

(b) Skeleton result of (a)

FIGURE 5.9 The result of a binary skeleton operation: (a) the original image, and (b) its skeleton result.

$$K = \max_{k}\{A \Theta kE \neq \emptyset\} \qquad (5.17)$$

The skeleton of an image A created by a structuring element E can be defined by means of the set operations:

$$S(A) = \bigcup_{k=0}^{K}\{(A\Theta kE) - ((A\Theta kE) \circ E)\} \qquad (5.18)$$

where K is given by Equation 5.17. Figure 5.9 shows the typical result of a binary skeleton operation.

5.3 MORPHOLOGY FOR GREY-SCALE IMAGES

Binary morphology can be extended to grey-scale images. In mathematical morphological methods, the difference between a binary image and a grey-scale image is that the former is described by its object set and its corresponding background set, and the latter is defined by the intensity $f(i, j)$ of each pixel point (i, j). Thus, the result of applying a morphological operation to a grey-scale image is a new image in which the intensity of each pixel is computed by using the respective morphological formulas.

5.3.1 Basic Grey-Scale Morphological Operations

5.3.1.1 Dilation Operation

The dilation of a grey-scale image $f(i, j), 0 \leq i \leq n_0 - 1, 0 \leq j \leq m_0 - 1$, and the structuring element $h(s, t), m_1 \leq s \leq m_2, n_1 \leq t \leq n_2$, denoted by $f \oplus h$ is defined [7] as

$$f \oplus h(i,j) = \max\{f(i-s, j-t) + h(s,t) \mid n_1 \leq s \leq n_2, 0 \leq (i-s) \leq n_0 - 1,$$
$$m_1 \leq t \leq m_2, 0 \leq (j-t) \leq m_0 - 1\}$$
(5.19)

where $0 \leq i \leq n_0 - 1, 0 \leq j \leq m_0 - 1$.

The structuring element in grey-scale morphology is similar to the convolution kernel in the convolution described in Chapter 2, Section 2.1.2. The following example uses the dilation of a one-dimensional signal to illustrate the operations involved in grey-scale dilation.

Example 5.6 $f(t)$ is a one-dimensional signal defined as follows:

$$f(t) = \{f(0), f(1), f(2), f(3), f(4), f(5), f(6), f(7), f(8), f(9)\}$$
$$= \{3, 5, 8, 4, 2, 6, 8, 10, 5, 4\}$$

The following short signal is used as a structuring element:

$$h(t) = \{h(-1), h(0), h(1)\} = \{1, 1, 1\}$$

Compute the new signal generated by the dilation $f \oplus h$.

Solution: The first two values of the new signal generated by the dilation $f \oplus h$ are computed as follows:

$$f \oplus h(0) = \max\{h(-1) + f(0+1), h(0) + f(0), h(1) + f(-1)\}$$
$$= \max\{h(-1) + f(1), h(0) + f(0)\}$$
$$= 6$$

$$f \oplus h(1) = \max\{h(-1) + f(1+1), h(0) + f(1), h(1) + f(1-1)\}$$
$$= \max\{h(-1) + f(2), h(0) + f(1), h(1) + f(0)\}$$
$$= 9$$

Similarly, the other values can be computed as follows:

$$f \oplus h(2) = 9, f \oplus h(3) = 9, f \oplus h(4) = 7, f \oplus h(5) = 9,$$
$$f \oplus h(6) = 11, f \oplus h(7) = 11, f \oplus h(8) = 11, f \oplus h(9) = 6$$

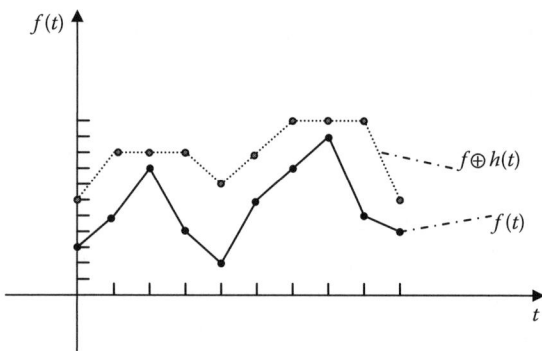

FIGURE 5.10   An illustration of $f \oplus h(t)$.

Figure 5.10 shows the relation between the given signal f, and its dilation by h. ∎

Example 5.7 The following matrix defines an 8-bit grey-scale image with the size 8×8.

$$f = \begin{bmatrix} f(0,0) & f(0,1) & \cdots & f(0,7) \\ f(1,0) & f(1,1) & \cdots & f(1,7) \\ \cdots & \cdots & \cdots & \cdots \\ f(7,0) & f(7,1) & \cdots & f(7,7) \end{bmatrix}$$

$$= \begin{bmatrix} 200 & 201 & 202 & 202 & 203 & 202 & 200 & 198 \\ 202 & 203 & 205 & 204 & 204 & 202 & 200 & 197 \\ 205 & 210 & 211 & 212 & 210 & 209 & 208 & 205 \\ 205 & 208 & 210 & 212 & 214 & 210 & 211 & 208 \\ 210 & 212 & 215 & 218 & 217 & 219 & 220 & 218 \\ 212 & 214 & 218 & 220 & 220 & 219 & 218 & 218 \\ 210 & 212 & 213 & 215 & 216 & 216 & 210 & 212 \\ 208 & 208 & 210 & 211 & 212 & 214 & 210 & 210 \end{bmatrix}$$

The following 3×3 matrix is used to construct a structuring element:

$$h = \begin{bmatrix} h(-1,-1) & h(-1,0) & h(-1,1) \\ h(0,-1) & h(0,0) & h(0,1) \\ h(1,-1) & h(1,0) & h(1,1) \end{bmatrix} = \begin{bmatrix} 0 & 1 & 0 \\ 1 & 1 & 1 \\ 0 & 1 & 0 \end{bmatrix}$$

Compute the intensities of the pixels located at (0,0) and (3,2) in the resulting image after performing the dilation $f \oplus h$.

Solution: The intensity of the pixel located at (0,0) in the image resulting from the dilation $f \oplus h$ is computed as follows:

$$f \oplus h(0,0)$$

$$= \max\{f(0-s, 0-t) + h(s,t) \mid s \in (-1,0,1) \cap (i-s) \geq 0; t \in (-1,0,1) \cap (j-t) \geq 0\}$$

$$= \max\{h(-1,-1) + f(1,1), h(-1,0) + f(1,0), h(0,-1) + f(0,1), h(0,0) + f(0,0)\}$$

$$= \max\{203, 203, 202, 201\}$$

$$= 203$$

The intensity of the pixel located at (3,2) in the image resulting from the dilation $f \oplus h$ is computed as follows:

$$f \oplus h(3,2) = \max \begin{cases} h(-1,-1) + f(3+1, 2+1), & h(-1,0) + f(3+1, 2), & h(-1,1) + f(3+1, 2-1), \\ h(0,-1) + f(3, 2+1), & h(0,0) + f(3,2), & h(0,1) + f(3, 2-1), \\ h(1,-1) + f(3-1, 2+1), & h(1,0) + f(3-1, 2), & h(1,1) + f(3-1, 2-1) \end{cases}$$

$$= \max \begin{bmatrix} 218, & 216, & 212, \\ 213, & 211, & 209, \\ 212, & 212, & 210 \end{bmatrix} = 218 \quad \blacksquare$$

Algorithm 5.3: Grey-scale dilation algorithm
For the given 8-bit grey-scale image $f(i, j)$ $0 \leq i, j \leq n-1$ with the given structure element matrix $e(s, t)$ $0 \leq s, t \leq m-1$:-
For $i = 0$ to $n-1$
For $j = 0$ to $n-1$
    $g(i, j) = f(i, j)$; // initialise the result image;
End-for
For $i = m$ to $n - m - 1$
    For $j = m$ to $n - m - 1$ do //exclude border rows and columns;
        $g(i, j) = 255$; max $= f(i, j)$;
        For $s = -m$ to m
            For $t = -m$ to m do
                temp $= f(i - s, j - t) + e(s, t)$;
                If (*temp* > max) then max = *temp*; End-If

End-For
   $g(i, j) = max$;
   If $((g(i, j) > 255))$ then $g(i, j) = 255$; End-If
End-For
Output of the resulting image: $g(i, j)$, $0 \leq i, j \leq n-1$
End-Algorithm

5.3.1.2 Erosion Operation

The erosion of a grey-scale image $f(i, j)$, $0 \leq i \leq n_0-1$, $0 \leq j \leq m_0-1$, and the structuring element $h(s, t)$, $m_1 \leq s \leq m_2$, $n_1 \leq t \leq n_2$, defined by the set operation $f \ominus h$ can be obtained by the following formula:

$$f \ominus h(i, j) = \min\{f(i+s, j+t) - h(s,t) | n_1 \leq s \leq n_2, 0 \leq (i+s) \leq n_0 - 1,$$
$$m_1 \leq t \leq m_2, 0 \leq (j+t) \leq m_0 - 1\} \tag{5.20}$$

where $0 \leq i \leq n_0-1$, $0 \leq j \leq m_0-1$.

The grey-scale dilation is similar to the convolution operation, whereas the grey-scale erosion is similar to the correlation operation defined in Chapter 2, Section 2.1.

Example 5.8 The one-dimensional signal f and the structuring element h as given in Example 5.6 are used here. Compute the new signal generated by the erosion $f \ominus h$.

Solution: The first two values of the new signal generated by the erosion $f \ominus h$ are computed as follows:

$$f \ominus h(0) = \min\{f(0-1) - h(-1), f(0) - h(0), f(0+1) - h(1)\}$$
$$= \min\{f(0) - h(0), f(1) - h(1)\}$$
$$= 2$$

$$f \ominus h(1) = \min\{f(1-1) - h(-1), f(1) - h(0), f(1+1) - h(1)\}$$
$$= \min\{f(0) - h(-1), f(1) - h(0), f(2) - h(1)\}$$
$$= 2$$

Similarly, the other values can be computed as follows:

$$f \ominus h(2) = 3, f \ominus h(3) = 1, f \ominus h(4) = 1, f \ominus h(5) = 1$$
$$f \ominus h(6) = 5, f \ominus h(7) = 4, f \ominus h(8) = 3, f \ominus h(9) = 3$$

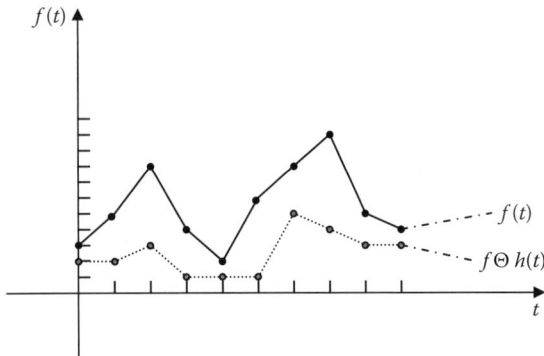

FIGURE 5.11 An illustration of $f \ominus h(t)$.

Figure 5.11 shows the given signal f and its erosion result obtained by $f\ominus h$.

Algorithm 5.4: Grey-scale erosion algorithm
For a given 8-bit grey-scale image $f(i, j)$, $0 \leq i, j \leq n - 1$
and a given structure element array $e(s, t)$, $0 \leq s, t \leq m - 1$:-
For $i = 0$ to $n - 1$
For $j = 0$ to $n - 1$ do {{$g(i, j) = f(i, j)$}}; // initialise the result image;
For $i = m$ to $n - m - 1$
For $j = m$ to $n - m - 1$ do
        min = 255;
        For $s = -m$ to m
        For $t = m$ to m do
            $temp = f(i + s, j + t) - e(s, t)$;
            If ($temp < min$) then $min = temp$; End-If
        End-for
        End-For
        $g(i, j) = min$; If ($g(i, j) < 0$) then $g(i, j) = 0$;
End-For
End-For
Output of the resulting image: $g(i, j)$, $0 \leq i, j \leq n - 1$
End-Algorithm

Figure 5.12 shows the typical result of grey-scale dilation and erosion operations.

5.3.2 Applications of Grey-Scale Morphological Operations

Although the formulas for grey-scale dilation and erosion provided by Equations 5.19 and 5.20 are different from that for binary images, the definitions of other grey-scale morphological operations are similar to binary

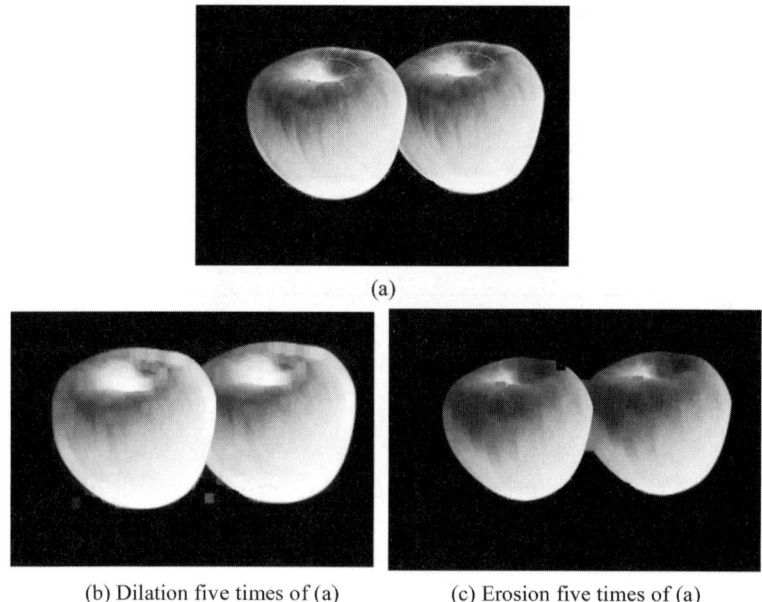

(a)

(b) Dilation five times of (a)   (c) Erosion five times of (a)

FIGURE 5.12   An example of grey-scale dilation and erosion.

morphological operations. For example, the grey-scale opening and closing operations are similar to the binary opening and closing ones defined in Equations 5.9 and 5.10. The opening $f \circ h$ and closing $f \bullet h$ of a grey-scale image f by a structuring element h are defined as follows:

$$f \circ h = (f \ominus h) \oplus h \qquad (5.21)$$

$$f \bullet h = (f \oplus h) \ominus h \qquad (5.22)$$

5.4   FURTHER READING

Mathematical morphology was proposed by Matheron and Serra in the late 1960s [1]. It is an efficient mathematical tool in image processing, especially in processing binary images. It can be used in edge detection and segmentation [8,9], shape recognition [10–12], texture analysis, and feature extraction [13–15]. In recent years, mathematical morphology has developed with more sophisticated methods, such as adaptive morphology [16,17], hierarchical morphology [8,18], and heterogeneous morphological granulometries [19].

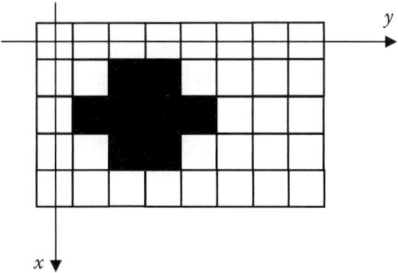

FIGURE 5.13   Q.2.

5.5   EXERCISES

Q.1   Prove that binary dilation operation is commutative and associative.

Q.2   Perform opening and closing operations to the binary image given in Figure 5.13 with the structuring element $E_3 = \{(-1,0), (0, 0), (1, 0)\}$.

Q.3   Prove that binary opening and closing operations are idempotent, that is,

$$(A \circ E) \circ E = A \circ E$$

$$(A \bullet E) \bullet E = A \bullet E$$

Q.4   Perform grey-scale opening using the signal and the structuring element defined in Example 5.6.

5.6   REFERENCES

1. J. Serra, *Image Analysis and Mathematical Morphology*, Academic Press, 1982.
2. M. Sonka, V. Hlavac, and R. Boyle, *Image Processing, Analysis and Machine Vision*, 2nd Edition, Thomson Learning and PPTPH, 1998.
3. K. Hrbacek and T. Jech, *Introduction to Set Theory*, 3rd Edition, CRC press, 1999.
4. R. Qiuqi, *Digital Image Processing Science*, Publishing House of Electronic Industry, 2001 (in Chinese).
5. E. Dougherty (Ed.), *Mathematical morphology in image processing*, M. Dekker, New York, 1993.
6. Z. S. G. Tari, J. Shah, and H. Pien, Extraction of shape skeletons from grey-scale images, *Computer Vision and Image Understanding*, Vol. 66(2): 133–146, 1997.

7. R. C. Gonzales and R. E. Woods, *Digital Image Processing*, Addison-Wesley, Reading, 1992.
8. G. Kukielka and J. Woznicki, Hierarchical method of digital image segmentation using multidimensional mathematical morphology, *Computer Analysis of Images and Patterns, Lecture Notes in Computer Science,* Vol. 2124: 581–588, 2001.
9. S. Beucher, Segmentation tools in mathematical morphology, Image Algebra and Morphological Image Processing, *Proceedings of SPIE,* Vol. 1350: 70–84, 1990.
10. D. Zhao and D. G. Daut, Morphological hit-or-miss transformation for shape recognition, *Journal of Visual Communication and Image Representation,* Vol. 2(3): 230–243, 1991.
11. C. C. Pu and F. Y. Shih, Morphological shape description and shape recognition using geometric spectrum on multidimensional binary images, *Industrial Electronics, Control, and Instrumentation, Proceedings of the IECON apos,* pp. 1371–1376, 1993.
12. W. Tong and S. Tatsumi, High-accuracy shape decomposition based on mathematical morphology in discrete circle space, *Systems and Computers in Japan,* Vol. 33(13): 85–95, 2002.
13. J. A. Bangham and S. Marshall, Image and signal processing with mathematical morphology, *Electronics and Communication Engineering Journal,* Vol. 10(3): 117–128, 1998.
14. X. Jin and C. H. Davis, New applications for mathematical morphology in urban feature extraction from high-resolution satellite imagery, *Applications of Digital Image Processing, Proceedings of the SPIE,* Vol. 5558: 137–148, 2004.
15. J. B. Mena, State of the art on automatic road extraction for GIS update: a novel classification, *Pattern Recognition Letters,* Vol. 24(16): 3037–3058, 2003.
16. F. Cheng and A. N. Venetsanopoulos, Adaptive morphological operators, fast algorithms and their application, *Pattern Recognition,* Vol. 33(6): 917–933, 2000.
17. O. Cuisenaire, Locally adaptable mathematical morphology using distance transformations, *Pattern Recognition,* Vol. 39(3): 405–416, 2006.
18. S. C. Pei and F. C. Chen, Hierarchical image representation by mathematical morphological subband decomposition, *Pattern Recognition Letters,* Vol. 16(2): 183–192, 1995.
19. S. Batman, E. R. Dougherty, and F. Sand, Heterogeneous morphological granulometries, *Pattern Recognition,* Vol. 33(6): 1047–1057, 2002.

5.7 PARTIAL CODE EXAMPLES

Project 5-1: Binary Erosion

(These codes can be found in CD: Project5-1\source code\project5-1View .cpp and morph.cpp)

```
#include "stdafx.h"
#include "project5_1.h"
#include "DlgMorph.h"
```

```c
#include "morph.h"
#include "project5_1Doc.h"
#include "project5_1View.h"
#ifdef _DEBUG
#define new DEBUG_NEW
#undef THIS_FILE
static char THIS_FILE[] = __FILE__;
#endif
/*************************************************
*********
* Function name:
* OnErosion()
*
* Parameter:
* None
*
* Return Value:
* None
*
* Description:
* Erosion
*
*************************************************
*******/
void CProject5_1View::OnErosion()
{
        // Get the document
        CProject5_1Doc* pDoc = GetDocument();
        ASSERT_VALID(pDoc);
        if(pDoc->m_hDIB == NULL)
                return ;
        LPSTR lpDIB = (LPSTR) ::GlobalLock((HGLOBAL) pDoc->m_hDIB);
        LPSTR lpDIBBits=::FindDIBBits (lpDIB);
        int cxDIB = (int) ::DIBWidth(lpDIB); // Size of DIB - x
        int cyDIB = (int) ::DIBHeight(lpDIB); // Size of DIB - y
        long lLineBytes = WIDTHBYTES(cxDIB * 8);
 // count the number of byte of the image per line

        int nMode;
```

```cpp
        // create the dialogue box
        DlgMorph dlgPara;

        // initialise the variable
        dlgPara.m_nMode = 0;

        // show the dialogue box to set the erosion direction
        if (dlgPara.DoModal() != IDOK)
        {
                return;
        }

        // get the erosion direction defined by the user
        nMode = dlgPara.m_nMode;
        int structure[3][3];
        if (nMode == 2)
        {
                structure[0][0]=dlgPara.m_nStructure1;
                structure[0][1]=dlgPara.m_nStructure2;
                structure[0][2]=dlgPara.m_nStructure3;
                structure[1][0]=dlgPara.m_nStructure4;
                structure[1][1]=dlgPara.m_nStructure5;
                structure[1][2]=dlgPara.m_nStructure6;
                structure[2][0]=dlgPara.m_nStructure7;
                structure[2][1]=dlgPara.m_nStructure8;
                structure[2][2]=dlgPara.m_nStructure9;
        }

        // delete the dialogue box
        delete dlgPara;

        // change the style of the cursor
        BeginWaitCursor();

        // call the function ErosionDIB()
        if (ErosionDIB(lpDIBBits, WIDTHBYTES(::DIBWidth(lpDIB) * 8), ::DIBHeight(lpDIB), nMode , structure))
        {
                // set the modification tag
```

```
                pDoc->SetModifiedFlag(TRUE);
                // update the view
                pDoc->UpdateAllViews(NULL);
        }
        else
        {
                // show the message to the user
                MessageBox(" failure to allocate the memory
or the intensity is not equal to 0 or 255!", "the system
show" , MB_ICONINFORMATION | MB_OK);
        }

        // unlock
        ::GlobalUnlock((HGLOBAL) pDoc->GetHDIB());
        // restore the style of the cursor
        EndWaitCursor();
}
#include "stdafx.h"
#include "morph.h"
#include "DIBAPI.h"
#include <math.h>
#include <direct.h>
/***********************************************************
 *
 * function name:
 * ErosiontionDIB()
 *
 * parameters:
 * LPSTR lpDIBBits - the pointer pointing to the origi-
nal image DIB
 * LONG lWidth - the width of the original image
 * (number of the pixels, 4 times)
 * LONG lHeight - the height of the original image
(pixel numbers)
 * int nMode      - erosion direction, 0- horizontal
direction,
 * 1- vertical direction, 2- user defined direction
 *     int structure[3][3]
                                    - user defined 3×3
structuring element matrix
 *
```

```
 * return value:
 * BOOL - return TRUE if success or return FALSE。
 *
 * Description:
 * Used to perform the erosion for the image. The structuring element matrix is
 * 3 pixel points in the horizontal direction, vertical direction or 3 by 3 points
 * defined by user
 *
 * the intensity of the pixel in the image should be 0 or 255.
 ***********************************************************
 ****************/
BOOL WINAPI ErosionDIB(LPSTR lpDIBBits, LONG lWidth, LONG lHeight, int nMode , int structure[3][3])
{
        // the pointer pointing to the original image
        LPSTR lpSrc;

        // the pointer pointing to buffer image
        LPSTR lpDst;

        // the pointer pointing to the buffer DIB image
        LPSTR lpNewDIBBits;
        HLOCAL hNewDIBBits;
        // cyclic variables
        long i;
        long j;
        int n;
        int m;
        // pixel intensity
        unsigned char pixel;
        // allocate the memory to save the new image temporary
        hNewDIBBits = LocalAlloc(LHND, lWidth * lHeight);
        if (hNewDIBBits == NULL)
        {
                // failure to allocate the memory
                return FALSE;
        }
```

```
        // lock the memory
        lpNewDIBBits = (char * )LocalLock(hNewDIBBits);
        // initialise the new memory to 255
        lpDst = (char *)lpNewDIBBits;
        memset(lpDst, (BYTE)255, lWidth * lHeight);
        if (nMode == 0)
        {
                // erosion in horizontal direction
                for(j = 0; j <lHeight; j++)
                {
                        for(i = 1;i <lWidth-1; i++)
                        {
                                // don't process the left and right border pixels
                                // in order not to over the borders
                                // the pointer pointing to the ith pixel of the jth row
                                //of the original image from the bottom
                                lpSrc = (char *)lpDIBBits + lWidth * j + i;
                                // the pointer pointing to the ith pixel of the jth row
                                //of the destination image from the bottom
                                lpDst = (char *)lpNewDIBBits + lWidth * j + i;
                                //get the pixel intensity of the pointer
                                pixel = (unsigned char)*lpSrc;
                                // the pixel intensity is not equal to 0 or 255
                                if (pixel != 255 && *lpSrc != 0)
                                    return FALSE;

                                // initialise the destination image to black
                                *lpDst = (unsigned char)0;
                                // if the current point or either of its horizontal neighbours in the
                                // original image is white,
                                // set the current point in the destination as white
                                for (n = 0;n < 3;n++ )
                                {
```

```
                                pixel = *(lpSrc+n-1);
                                if (pixel == 255 )
                                {
                                        *lpDst = (unsigned
char)255;
                                        break;
                                }
                        }

                }
            }
        }
        else if (nMode == 1)
        {
                // erosion in vertical direction
                for (j = 1; j <lHeight-1; j++)
                {
                        for (i = 0;i <lWidth; i++)
                        {
                                // don't process the top and
bottom border pixels
// in order not to over the borders
                                // the pointer pointing to the
ith pixel of the jth row
//of the original image from the bottom
                                lpSrc = (char *)lpDIBBits +
lWidth * j + i;
// the pointer pointing to the ith pixel of the jth row
//of the destination image from the bottom
                                lpDst = (char *)lpNewDIBBits +
lWidth * j + i;
                                //get the intensity of the
current pointer
                                pixel = (unsigned char)*lpSrc;
                                // the pixel intensity is not
equal to 0 or 255
                                if (pixel != 255 && *lpSrc != 0)
                                        return FALSE;
                                //initialise the destination
image to black
                                *lpDst = (unsigned char)0;
                                // if the current point or
either of its vertical neighbours in the
```

```
                                // original image is white, 
                                // set the current point in the destination as white
                                for (n = 0;n < 3;n++ )
                                {
                                        pixel = *(lpSrc+(n-1)*lWidth);
                                        if (pixel == 255 )
                                        {
                                                *lpDst = (unsigned char)255;
                                                break;
                                        }
                                }

                        }
                }
        }
        else
        {
                // erosion with user defined structuring element matrix
                for ( j = 1; j <lHeight-1; j++)
                {
                        for(i = 0;i <lWidth; i++)
                        {
        // don't process the border pixels
        // in order not to over the borders
                                // the pointer pointing to the ith pixel of the jth row
                                //of the original image from the bottom
                                lpSrc = (char *)lpDIBBits + lWidth * j + i;
                                // the pointer pointing to the ith pixel of the jth row
                                //of the destination image from the bottom
                                lpDst = (char *)lpNewDIBBits + lWidth * j + i;
                                // get the intensity of the current pointer
                                pixel = (unsigned char)*lpSrc;
                                // the intensity is not equal to 0 or 255
```

```
                if(pixel != 255 && *lpSrc != 0)
                    return FALSE;
                // initialise the destination image to black
                *lpDst = (unsigned char)0;
                // if one of the neighbours of the current pixel corresponding
                //to the structuring element in the original image is white,
                // set the current point in the destination as white
                // note that the content in the DIB image is from bottom to top
                for (m = 0;m < 3;m++ )
                {
                    for (n = 0;n < 3;n++)
                    {
                        if(structure[m][n] == -1)
                            continue;
                        pixel = *(lpSrc + ((2-m)-1)*lWidth + (n-1));
                        if (pixel == 255 )
                        {
                            *lpDst = (unsigned char)255;
                            break;
                        }
                    }
                }
            }
        }
    }
    // copy the result image of erosion
    memcpy(lpDIBBits, lpNewDIBBits, lWidth * lHeight);
    // release the memory
    LocalUnlock(hNewDIBBits);
    LocalFree(hNewDIBBits);
    // return
    return TRUE;
}
```

Project 5-2: Binary Skeleton

(These codes can be found in CD: Project5-1\source code\project5-2 View.cpp)

```
#include "stdafx.h"
#include "project5_2.h"
#include "morph.h"
#include "project5_2Doc.h"
#include "project5_2View.h"
#ifdef _DEBUG
#define new DEBUG_NEW
#undef THIS_FILE
static char THIS_FILE[] = __FILE__;
#endif
/*************************************************
*********
* Function name:
* Onskeletonisation()
*
* Parameter:
* None
*
* Return Value:
* None
*
* Description:
* skeletonisation
*
*************************************************
*******/
void CProject5_2View::Onskeletonisation()
{
    // get the document
    CProject5_2Doc* pDoc = GetDocument();
    ASSERT_VALID(pDoc);
    if(pDoc->m_hDIB == NULL)
        return ;
    LPSTR lpDIB = (LPSTR) ::GlobalLock((HGLOBAL) pDoc->m_hDIB);
    LPSTR lpDIBBits=::FindDIBBits (lpDIB);
    int cxDIB = (int) ::DIBWidth(lpDIB); // Size of DIB - x
```

```cpp
        int cyDIB = (int) ::DIBHeight(lpDIB); // Size of
DIB - y
        long lLineBytes = WIDTHBYTES(cxDIB * 8);
// count the the number of bytes of the image per line
        // change the style of the cursor
        BeginWaitCursor();
        // cyclic variables
        int i;
        int j;
        int k;
        int lWidth = cxDIB;
        int lHeight = cyDIB;
        // the pointer pointing to the buffer image
        unsigned char *lpSrc;
        unsigned char *lpDst;
        // the pointer pointing to the buffer DIB image
        LPSTR lpNewDIBBits1,lpNewDIBBits2;
        HLOCAL hNewDIBBits1,hNewDIBBits2;
        // allocate the memory to save the new image tem-
porary
        hNewDIBBits1 = LocalAlloc(LHND, lWidth * lHeight);
        hNewDIBBits2 = LocalAlloc(LHND, lWidth * lHeight);

        // lock the memory
        lpNewDIBBits1 = (char * )LocalLock(hNewDIBBits1);
        lpNewDIBBits2 = (char * )LocalLock(hNewDIBBits2);
        // initialise the new memory to 0
        lpSrc = (unsigned char *)lpNewDIBBits1;
        memset(lpSrc, (BYTE)0, lWidth * lHeight);
        lpDst = (unsigned char *)lpNewDIBBits2;
        memset(lpDst, (BYTE)0, lWidth * lHeight);
        int nCount=0,m_nSEWidth=4;
        unsigned char** pBufSK=new unsigned char*[m_
nSEWidth];          for(j=0;j<m_nSEWidth;j++)
        {
                pBufSK[j]=new unsigned char [lWidth*lHeight];
                memset (pBufSK[j],0,lWidth*lHeight);
        }
        unsigned char* pDest=new unsigned
char[lWidth*lHeight];
        memset(pDest,0,lWidth*lHeight);
```

```
        while(nCount<m_nSEWidth)
        {
                nCount++;
                i=0;
                memcpy(lpNewDIBBits1, lpDIBBits, lWidth *
lHeight);
                while(i++<nCount)
                {
                        ErosionDIB(lpNewDIBBits1, lWidth,
lHeight);
                }
                memcpy(lpNewDIBBits2, lpNewDIBBits1, lWidth
* lHeight);
                OpenDIB(lpNewDIBBits2, lWidth, lHeight);

 for(i=0;i<lHeight;i++)
                {
                        for(j=0;j<lWidth;j++)
                        {
                        lpSrc = (unsigned char*)lpNewDIBBits1
+ lWidth * i + j;
                        lpDst = (unsigned char*)lpNewDIBBits2
+ lWidth * i + j;
                                if((*lpSrc==0)&&(*lpDst==255))
                                    pBufSK[nCount-1]
[i*lWidth+j]=0;
                                else
if((*lpSrc==255)&&(*lpDst==0))
                                    pBufSK[nCount-1]
[i*lWidth+j]=0;
                                else
                                    pBufSK[nCount-1]
[i*lWidth+j]=255;
                        }
                    }
            }
 for(k=0;k<m_nSEWidth-1;k++)
        {
                for(i=0;i<lHeight;i++)
                {
                        for(j=0;j<lWidth;j++)
```

```
                    {
                        if((pBufSK[k][i*lWidth+j]==0)
||(pBufSK[k+1][i*lWidth+j]==0))
                            pBufSK[k+1]
[i*lWidth+j]=0;
                        else
                            pBufSK[k+1]
[i*lWidth+j]=255;
                    }
                }
            }
    memcpy(lpDIBBits,pBufSK[k],lWidth*lHeight);
    // release the memory
    LocalUnlock(hNewDIBBits1);
    LocalFree(hNewDIBBits1);
    LocalUnlock(hNewDIBBits2);
    LocalFree(hNewDIBBits2);
    delete[] pDest;
    for(j=0;j<m_nSEWidth;j++)
    {
        delete[] pBufSK[j];
    }
    delete[] pBufSK;
    // set the modification tag
    pDoc->SetModifiedFlag(TRUE);

    // update the view
    pDoc->UpdateAllViews(NULL);

    // unlock
    ::GlobalUnlock((HGLOBAL) pDoc->GetHDIB());
    // restore the style of the cursor
    EndWaitCursor();
}
```

CHAPTER 6

Image Compression

Storage and transmission are essential processes in image processing. As discussed in Chapter 1, images are generally stored in the bitmap format, and the memory in spatial dimensions could be very large if images are stored directly without preprocessing. For example, the data of an 8-bit grey-scale image with the resolution 256 × 256 requires a total memory of 65536 bytes (or 64 kilobytes). The memory required for a true colour image increases to 64 kilobytes × 3 = 192 kilobytes. Under the National Television Standard Committee (NTSC) standard, 30 frames of images are played in one second to ensure continuous vision effect. Suppose the images are true colour having a resolution of 720 × 576, the images played in one second would require the storage size of 720 × 576 × 3 × 30 = 37324800 bytes = 36 megabytes. Such a huge amount of data would cause enormous difficulties during storage or transmission. Therefore, compression of original images is inevitable to facilitate transmission or other processes.

The essence of compression is to use a compressed file with smaller storage size requirements to replace the original one. The compressed file can be reverted to the original file through decompression. If the decompressed image is identical to the original image, the corresponding compression method is called *lossless compression*; otherwise, it is called *lossy compression*. Common lossy compression methods include predictive compression, vector quantisation, transform encoding, wavelet compression, and fractal compression. The last two methods are considered as state-of-the-art transform compression techniques.

Compression rate can be used to assess the efficiency of a compression method. It is defined as the ratio of the size of the original file to the compressed file. If the size of the original file and the compressed file are

a bytes and b bytes, respectively, the compression rate is calculated as a/b. Another common method of calculating the compression rate is by counting the number of bits in the compressed file needed to represent a pixel in the original file, and it is written using the unit bpp (bits per pixel). For example, suppose the size of the original image is $m \times n$ and that of the compressed file is b bytes, the compression rate is $8b/(m \times n)$ bpp.

The difference between the decompressed and the original images in the case of lossy compression needs to be evaluated. The smaller the difference, the higher the quality of compression. Obviously, lossless compression has the best quality of compression. On the other hand, the compression rate of a lossy compression method is certainly related to the difference between the decompressed and the original images. In general, the higher the compression rate, the larger such difference would be.

In this chapter, some standards of image quality measurement are introduced first. Huffman encoding and runlength encoding are discussed in Sections 6.2.1 and 6.2.2. Prediction compression, vector quantisation along with wavelet compression and fractal compression in transform encoding are discussed in Section 6.3. Two common standards of image compression—Joint Photographic Experts Group (JPEG) and Moving Pictures Experts Group (MPEG)—are also introduced. The last section contains further readings and future research directions.

6.1 IMAGE FIDELITY METRICS

Although subjective assessment may be used to observe the extent of difference between a decompressed image and its original image, it is important to have objective measurements in order to quantify image fidelity. The root-mean-square (rms) error and peak-to-peak signal-to-noise ratio (PSNR) [1], among others, are two commonly used metrics.

Suppose, the intensity matrix of the original $m \times n$ image is $f(i, j)$, $i = 0, 1, \ldots, m-1$; $j = 0, 1, \ldots, n-1$ and that of the decompressed $m \times n$ image is $g(i, j)$, $i = 0, 1, \ldots, m-1$; $j = 0, 1, \ldots, n-1$, then the two metrics are defined as follows.

1. Root-mean-square (rms) error: The rms error of the decompressed image $g(i, j)$ and the original image $f(i, j)$ is defined as

$$rms = \sqrt{\frac{\sum_{i=0}^{m-1}\sum_{j=0}^{n-1}(g(i,j)-f(i,j))^2}{m \times n}} \quad (6.1)$$

(a) Original image   (b) PSNR = 31.59   (c) PSNR = 28.54

FIGURE 6.1   Images with different PSNR values.

2. Peak-to-peak signal-to-noise ratio (PSNR): PSNR represents the ratio of the maximum possible power of a signal and the possible power of the error. It is also called the quantisation noise ratio with the unit db (decibel). In the case of an 8-bit grey-scale image, PSNR of $g(i, j)$ and $f(i, j)$ is defined as

$$PSNR = 10 \times \log_{10} \frac{255^2}{\frac{1}{m \times n} \sum_{i=0}^{m-1} \sum_{j=0}^{n-1} (f(i,j) - g(i,j))^2} \quad (6.2)$$

Here, the denominator reflects the mean square error due to the difference between the original and decompressed images. There are other forms of PSNR where the denominator uses rms, and in this case, the leading constant is therefore 20. Some examples showing decompressed results with their corresponding PSNR values are shown in Figure 6.1.

6.2 LOSSLESS COMPRESSION

In Chapter 1, it is pointed out that .bmp file format uses runlength encoding. The JPEG compression standard requires the use of Huffman encoding or runlength encoding to process the resulting coefficients through transformations. The Huffman and runlength encodings are typical lossless compression methods, and are described in the following sections.

6.2.1 Huffman Encoding

Huffman encoding [2] is a statistical-theory-based encoding. Its main idea is to construct a shorter codeword for the source symbol with a higher occurring frequency, and a longer codeword for the source symbol with a lower occurring frequency. The Huffman encoding forms a codeword for each source symbol by constructing a binary Huffman tree.

A source symbol is used to refer to a certain intensity of a pixel that occurs in a given image. The first step of the Huffman encoding is to calculate the occurring probability of each source symbol. The source symbols are then arranged according to the decreasing order of probability. Assuming these source symbols as the leaves of a tree, the probability of every source symbol is considered as the weight of the corresponding node. A parent node is generated for the two least weighted nodes and has a weight equal to the sum of the weights of the two children. The process is repeated until the root of the tree occurs. Starting from the root, the code 0 or 1 is assigned to the two branches of each node of the tree. The codes from the root to every leaf source symbol forms a binary string that is the codeword of the corresponding source symbol. All codewords together form the code table.

The compression process uses codewords to replace the corresponding intensities of pixels (source symbol) in order to form a compressed file. To facilitate the decoding process, the code table should be included as part of the compressed file.

Example 6.1 Suppose $f(i, j)$ denotes the intensity matrix of a 6-bit 8×8 grey-scale image as given by the following matrix:

$$\begin{bmatrix} 15 & 20 & 15 & 15 & 20 & 25 & 15 & 20 \\ 30 & 25 & 25 & 30 & 20 & 20 & 30 & 30 \\ 35 & 30 & 35 & 40 & 40 & 30 & 25 & 30 \\ 40 & 35 & 40 & 50 & 50 & 40 & 35 & 35 \\ 40 & 40 & 50 & 45 & 45 & 55 & 50 & 40 \\ 45 & 50 & 55 & 50 & 55 & 60 & 55 & 45 \\ 40 & 45 & 40 & 45 & 50 & 60 & 45 & 40 \\ 30 & 35 & 25 & 25 & 30 & 35 & 30 & 20 \end{bmatrix}$$

Construct the Huffman codeword.
Solution: The source symbols in this case are

$$s_1 = 15, s_2 = 20, s_3 = 25, s_4 = 30, s_5 = 35, s_6 = 40, s_7 = 45,$$

$$s_8 = 50, s_9 = 55, s_{10} = 60$$

Using the intensity matrix, one can find the number of occurrences of each source symbol. Let c_i be the number of occurrences of s_i, $i = 1, 2, \ldots, 10$. The values of c_i are

$$c_1 = 4, c_2 = 6, c_3 = 6, c_4 = 10, c_5 = 7, c_6 = 11, c_7 = 7, c_8 = 7, c_9 = 4, c_{10} = 2$$

The probability p_i of the source symbol s_i, $i = 1, 2, \ldots, 10$, is given as follows:

$$p_1 = \frac{4}{64}, p_2 = \frac{6}{64}, p_3 = \frac{6}{64}, p_4 = \frac{10}{64}, p_5 = \frac{7}{64}, p_6 = \frac{11}{64},$$

$$p_7 = \frac{7}{64}, p_8 = \frac{7}{64}, p_9 = \frac{4}{64}, p_{10} = \frac{2}{64}$$

Now, the Huffman tree is ready to be constructed, and the process is described as follows:

1. Rearrange source symbols according to the descending order of probability:

$$p_6 > p_4 > p_5 = p_7 = p_8 > p_2 = p_3 > p_1 = p_9 > p_{10}$$

$$s_6, s_4, s_5, s_7, s_8, s_2, s_3, s_1, s_9, s_{10}$$

Let these source symbols be the leaves of the Huffman tree, and their corresponding probabilities be the weights of the leaves.

| 6 | 4 | 5 | 7 | 8 | 2 | 3 | 1 | 9 | 10 |

2. Generate a new node 11, which is the parent node of the two least weighted nodes 9 and 10. The weight p_{11} of the new node 11 is the sum of the weights of the two child nodes, that is, $p_{11} = p_9 + p_{10} = \frac{6}{64}$.

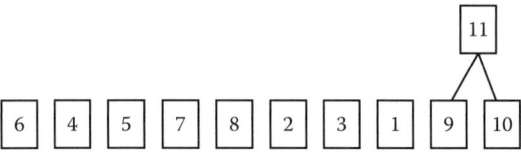

3. Arrange the new nodes according to the descending order of weights:

$$p_6 > p_4 > p_5 = p_7 = p_8 > p_2 = p_3 > p_{11} = p_1$$

4. Repeat steps 2 and 3 until the root of the tree occurs. Figure 6.2 depicts the tree.

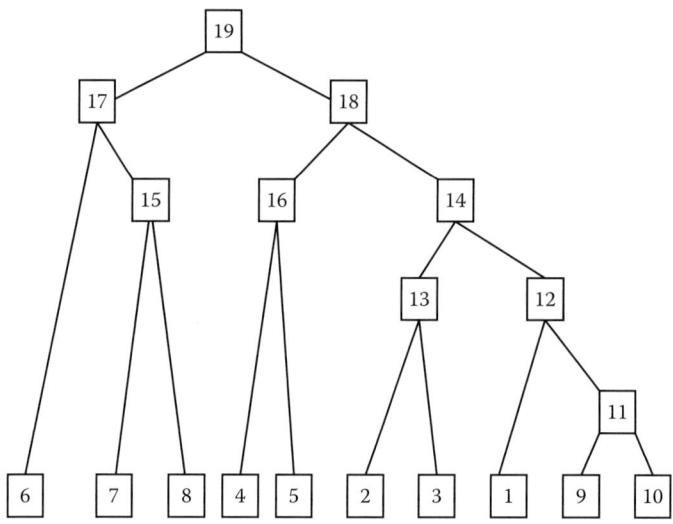

FIGURE 6.2 Huffman tree of Example 6.1.

5. Assign the code 0 or 1 to the two branches of every node of the tree. Figure 6.3 depicts the result.

6. The codeword for each source symbol is formed by taking the binary string from the root to the corresponding leaf:

$s_6 : 00;$

$s_7 : 010;$ $s_8 : 011;$ $s_4 : 100;$ $s_5 : 101$

$s_2 : 1100;$ $s_3 : 1101;$ $s_1 : 1110$

$s_9 : 11110;$ $s_{10} : 11111$ ∎

Note that the codeword obtained from Huffman encoding has this unique prefix property—no codeword is a prefix to any other codewords, and each codeword is unambiguous. The Huffman encoding algorithm for a grey-scale image is described here.

Algorithm 6.1: The Huffman coding algorithm
Given the image $f(i, j) : 0 \leq i \leq m-1, 0 \leq j \leq n-1$;
Let L be the grey levels of the image.
Initialise L word_nodes each representing a grey level:
    word_node (l). word $:= l;$

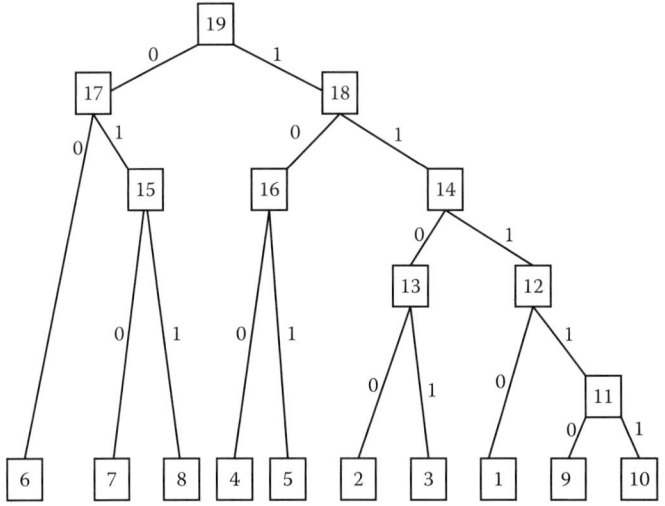

FIGURE 6.3 Assign a binary number to every branch of the Huffman tree given in Figure 6.2.

    word_node (*l*).weight := the number of occurrences of *l* as the pixel values in the image;

Define a node of Huffmantree H_node with five components:

    H_node.word, H_node.weight, H_node.parent, H_node.lchild, H_node.rchild;

Initialise a Huffman tree *T* with $K = 2^L - 1$ H_nodes: // construct Huffman Tree

For $j = 1$ to L do

    $T(j)$.word = word_node (j). word;

    $T(j)$.weight = word_node (j). p;

    $T(j)$.parent = 0;

    $T(j)$.lchild= 0;

    $T(j)$.rchild = 0;

End-For

For $j = L + 1$ to K do

    Find two H_nodes with minimal weights in *T*:

        $T(j_1)$, $j_1 < j$; and $T(j_2)$, $j_2 < j$

    Assign component values for the new H_node $T(j)$:

        $T(j)$.word = 0;

$T(j)$.weight $= T(j_1)$.weight $+ T(j_2)$.weight;
$T(j)$.parent $= 0$;
$T(j)$.lchild $= j_1$;
$T(j)$.rchild $= j_2$;
  Change the parent of two child H_nodes:
    $T(j_1)$.parent $= j$; $T(j_2)$.parent $= j$;
End-For // End of Tree Construction
For $i = 1$ to L do // assign a binary string for each word
    $c = i$; code_word(code_word(i)):= ''; //empty string
    while $c \neq 0$ do
    {   c_parent $= T(c)$.parent;
         if (T[c_parent].lchild==c)
                code_word(i) = '0' + code_word(i)
         else
                code_word(i) = '1' + code_word(i)
         endif
         $c = T(c)$.parent
    }
End-For
Create the compression file *comp_file*;
// generate a compression file for the image
// Store the codeword table *comp_word* to *comp_file*;
For $i = 0$ to $n - 1$
For $j = 0$ to $m - 1$
         Store *comp_word* $(f(i, j))$ to *comp_file*;
End-For
End-Algorithm

6.2.2 Runlength Encoding

Runlength encoding is an easy-to-use coding method. Its main idea is to use a source symbol and its respective number of consecutive occurrences, instead of listing every occurrence of the same source symbol. For example, if the source file has the following data:

a a a a a a a a a a a a a a a a b b b b a a a a a a a a c c c c c c c

The encoded file by using the runlength method becomes

a 15 b 4 a 8 c 7

6.3 LOSSY COMPRESSION

Image data exhibits certain redundancy as far as human vision of the data is concerned. In other words, the removal of certain parts of the data information might not affect the overall effect of vision. For this reason, lossy compression is more commonly used in image compression. Nowadays, lossy compression methods are loosely classified as predictive encoding, vector quantisation, and transform encoding.

6.3.1 Predictive Compression Methods

In the frame of an image, the intensities of neighbouring pixels often have relatively close correlation. Therefore, the intensity of a pixel can be predicted by the one that occurs previously, and the predictive error is the removal of the actual intensity from the predicted intensity. Thus, the predictive error may be used to substitute the original intensity to be encoded and transmitted [3,4]. Usually, the range of the predictive error is much smaller than that of the original intensity of a pixel, which leads to shorter codewords and a higher compression ratio.

The most commonly used predictive compression method is differential pulse code modulation (DPCM), which is based on pulse code modulation (PCM) which converts analogue signals to digital signals through the processing steps of low-pass filtering, signal sampling, quantisation, and encoding into binary numbers, etc. [5]. Instead of processing analogue signals, in DPCM, the predictive error (i.e., the input signal of DPCM) is processed by using PCM. Figure 6.4 depicts the encoding and decoding principles of DPCM.

In the encoding process, let $f(i, j)$ be the original intensity of the pixel at location (i, j), and $\hat{f}(i, j)$ be an estimation of $f(i, j)$ according to the intensities of the neighbouring pixels of (i, j). The difference $e(i, j)$ between $f(i, j)$ and $\hat{f}(i, j)$ can be evaluated by

$$e(i,j) = f(i,j) - \hat{f}(i,j) \tag{6.3}$$

This error is sent to the encoder and predictor after quantisation. The output from the encoder forms the compressed file, and the output from the predictor is the intensity used in the prediction of the next pixel.

In the decoding part, the output $e'(i, j)$ from the decoder is an approximation of $e(i, j)$ obtained by quantisation, and an approximation of the original value $f(i, j)$ may be computed as $f'(i,j) = \hat{f}(i,j) + e'(i,j)$.

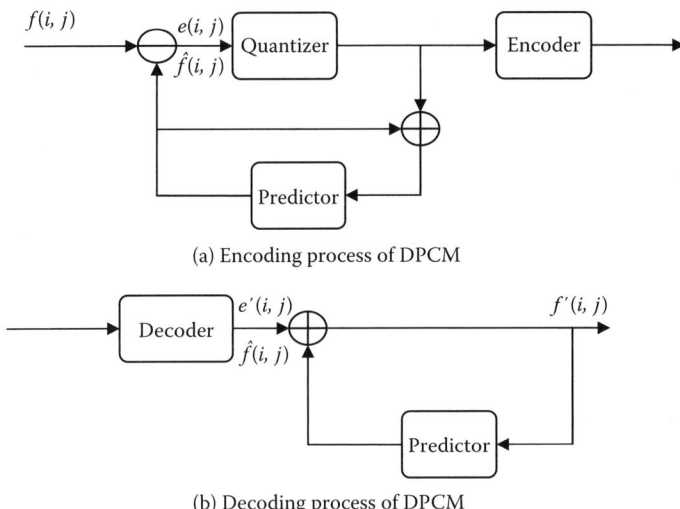

FIGURE 6.4 Encoding and decoding principles of DPCM. (a) Encoding process of DPCM and (b) decoding process of DPCM.

6.3.2 Vector Quantisation

Vector quantisation is the generalisation of scalar quantisation, as introduced in Chapter 1. In scalar quantisation, a range of real numbers are represented by a quantum value. Similarly, in vector quantisation, a set of vectors are represented by a quantum vector called the *code vector*.

The vector quantisation method in image compression is based on the principle of block coding. In order to describe block coding, the concept of a block partition of an image is needed to assist the description, and it is briefly described here. For convenience, suppose the given image is a square and the intensity matrix is denoted as $f(i,j), i=0,1,\ldots, 2^N-1; j=0,1,\ldots, 2^N-1$. The image is partitioned into nonoverlapping fixed size blocks [6]. In other words, the intensity matrix

$$\mathbf{P}_f = \begin{bmatrix} f(0,0) & f(0,1) & \cdots & f(0,2^N-1) \\ f(1,0) & f(1,1) & \cdots & f(1,2^N-1) \\ \vdots & & & \\ f(2^N-1,0) & f(2^N-1,1) & \cdots & f(2^N-1,2^N-1) \end{bmatrix} \quad (6.4)$$

is partitioned into intensity submatrices $\mathbf{R}_{s,t}, 0 \le s,t \le 2^{N-n}-1$, each having the size $2^n \times 2^n$:

$$\mathbf{P}_f = \begin{bmatrix} \mathbf{R}_{0,0} & \mathbf{R}_{0,1} & \cdots & \mathbf{R}_{0,2^{N-n}-1} \\ \mathbf{R}_{1,0} & \mathbf{R}_{1,1} & \cdots & \mathbf{R}_{1,2^{N-n}-1} \\ \vdots & & & \\ \mathbf{R}_{2^{N-n}-1,0} & \mathbf{R}_{2^{N-n}-1,1} & \cdots & \mathbf{R}_{2^{N-n}-1,2^{N-n}-1} \end{bmatrix} \quad (6.5)$$

where the submatrix $\mathbf{R}_{s,t}$ has the form

$$\mathbf{R}_{s,t} = \begin{bmatrix} f(2^n s, 2^n t) & f(2^n s, 2^n t+1) & \cdots & f(2^n s, 2^n t + 2^n - 1) \\ f(2^n s+1, 2^n t) & f(2^n s+1, 2^n t+1) & \cdots & f(2^n s+1, 2^n t + 2^n - 1) \\ \vdots & & & \\ f(2^n s + 2^n -1, 2^n t) & f(2^n s + 2^n -1, 2^n t+1) & \cdots & f(2^n s + 2^n -1, 2^n t + 2^n -1) \end{bmatrix}$$

(6.6)

The intensity submatrix $R_{s,t}$ is collocated using a row-wise data structure that leads to the source vector $X_{s,t}$. For example, when n is chosen as 2, the size of the matrix $R_{0,0}$ is 4×4, and the source vector $X_{0,0}$ has the form

$$X_{0,0} = (f(0,0), f(0,1), f(0,2), f(0,3), f(1,0), f(1,1), f(1,2), f(1,3),$$

$$f(2,0), f(2,1), f(2,2), f(2,3), f(3,0), f(3,1), f(3,2), f(3,3))$$

with its dimension being 16. In general, an image is represented by the set of source vectors $\aleph = \{X_{s,t}; 0 \leq s,t \leq 2^{N-n}-1\}$, each having the dimension $K = 2^{2n}$. $\aleph$ is collocated using a row-wise data structure, which leads to the set $\Im = \{X_1, X_2, ..., X_v\}$, where $v = 2^{2(N-n)}$.

The main idea of vector quantisation [7] is to construct a partition P of $\Im$:

$$P = \{\Im_1, \Im_2, ..., \Im_u\} \quad (6.7)$$

satisfying the conditions

$$\Im = \Im_1 \cup \Im_2 \cup ... \cup \Im_u$$

and

$$\Im_i \cap \Im_j = \emptyset, i \neq j$$

All the source vectors in the same subset $\Im_i$ are mapped to the same quantum vector C_i called a code vector, which has the same dimension K as the other source vectors of $\Im_i$. In other words, for any $X \in \Im_i$, the quantum vector is formed as follows:

$$Q(X) = C_i = (c_{i1}, c_{i2}, ..., c_{iK}) \quad (6.8)$$

where $X = (f_1, f_2, ..., f_K)$ contains elements selected from $f(i, j)$ using Equation 6.6. The code vectors created using Equation 6.8 form the elements of the codebook C, which is defined as

$$C = \{C_1, C_2, ..., C_u\} \quad (6.9)$$

During the process of decompression, code vectors can be used to approximate the original source vectors. The average error E_{ave} due to quantisation can be calculated by using a squared-error distortion:

$$E_{ave}(P,C) = \frac{1}{vK} \sum_{i=1}^{u} \sum_{X \in \Im_i} \|X - Q(X)\|^2$$

$$= \frac{1}{vK} \sum_{i=1}^{u} \sum_{X \in \Im_i} \sum_{k=1}^{K} (f_X - c_{ik})^2 \quad (6.10)$$

The vector quantisation method can now be described as an optimisation problem:

Given the set of K-dimensional source vectors $\Im = \{X_1, X_2, ..., X_v\}$ and the number of code vectors u, find a partition $P^* = \{\Im_1^*, \Im_2^*, ..., \Im_u^*\}$ and a codebook $C^* = \{C_1^*, C_2^*, ..., C_u^*\}$, which minimises the average error E_{ave} defined by Equation 6.10. Mathematically, this can be written as

$$E_{ave}(P^*, C^*) = \min_{P,C} E_{ave}(P, C) \quad (6.11)$$

The optimal design (P^*, C^*) should satisfy the following two criteria [8,9]:

1. The nearest neighbour condition:

$$\Im_i = \{X : \quad \|X - C_i\|^2 \leq \|X - C_j\|^2, j = 1, 2, ..., u\}, \text{ where } i = 1, 2, ..., u \quad (6.12)$$

2. The centroid condition:

$$C_i = \frac{\sum_{X \in \mathfrak{I}_i} X}{|\mathfrak{I}_i|} \tag{6.13}$$

where $|\mathfrak{I}_i|$ denotes the number of vectors in $\mathfrak{I}_i$ and $i = 1, 2, \ldots, u$.

The above criteria can be used to generate an optimal partition and the corresponding codebook by an iterative process. A commonly used iterative algorithm known as the (LBG) algorithm is summarised here for reference.

Algorithm 6.2: LBG design algorithm for image vector quantisation

Given the image $f(i,j): 0 \le i, j \le 2^N - 1$, the tolerance $\varepsilon \ll 1$, and the final number of code-vectors 2^h;

Prepare the set of source vectors: $\mathfrak{I} = \{X_1, X_2, \ldots, X_v\}$, $v = 2^{2(N-n)}$, where $X_i = (f_{i1}, f_{i2}, \ldots, f_{iK})$ with $K = 2^{2n}$.

$u = 1;$ // initial partition;

$C_1^{(0)} = \left(c_{11}^{(0)}, c_{12}^{(0)}, \ldots, c_{1K}^{(0)}\right) = \dfrac{1}{v} \sum\limits_{i=1}^{v} X_i;$ // initial codebook including one code-vector

$E_{ave}^{(0)} = \dfrac{1}{vK} \sum\limits_{m=1}^{v} \sum\limits_{j=1}^{K} \left(f_{mj} - c_{1j}^{(0)}\right)^2$ // average error

$j = 0;$ // iteration times

Do while ($j \le h$)

  $j = j + 1;$
  For $i = 1$ to u do   // splitting the codebook

    $C_i^{(j)} = (1+\varepsilon)C_i^{(j-1)};$

    $C_{u+i}^{(j)} = (1-\varepsilon)C_i^{(j-1)}$

  End-For;
  $u = 2 \times u;$ initialise partition $P = \{\mathfrak{I}_1, \mathfrak{I}_2, \ldots, \mathfrak{I}_u | \mathfrak{I}_i = \emptyset, 1 \le i \le u\}$
  Repeat {
    For $m = 1$ to v // repartition the source vectors
     Solve
     $\|X_m - C_{m^*}^{(j)}\| = \min\limits_{1 \le i \le u} \|X_m - C_i^{(j)}\|^2;$
     Let $X_m \in \mathfrak{I}_{m^*}$, i.e., $Q(X_m) = C_{m^*};$
  End-For;

For $i = 1$ to u do
// update the code vectors
$$C_i^{(j)} = \frac{1}{|\mathfrak{I}_i|} \sum_{X \in \mathfrak{I}_i} X$$
End-For;
Calculate current average error:
$$E_{ave}^{(j)} = \frac{1}{vK} \sum_{m=1}^{v} \| X_m - Q(X_m) \|^2$$
} Until $\left(\left(E_{ave}^{(j-1)} - E_{ave}^{j} \right) \big/ E_{ave}^{(j-1)} \le \varepsilon \right)$

End-Do.
Output the partition $P = \{\mathfrak{I}_1, \mathfrak{I}_2, ..., \mathfrak{I}_u\}$ and the codebook $C_1^{(j)}, C_2^{(j)}, ..., C_u^{(j)}$;
End-Algorithm

Note that in the compression file, the source vector X_m is represented by the index of the code vector $Q(X_m)$ in the codebook. In order to perform the decompression correctly, the codebook should be included in the compression file.

6.3.3 Wavelet Compression

In image processing, another commonly used lossy compression method is transform encoding. It does not work on the intensity of the pixel directly, but transforms the intensity of a pixel and encodes the result of the transformation. For example, the JPEG standard adopts the discrete cosine transform (DCT) introduced in Chapter 2, Section 2.3, and encodes the transformed coefficients afterwards. This section explains a typical transform encoding method—the wavelet image compression.

In Chapter 2, Equation 2.34 defines the wavelet transform of a signal $f(t)$

$$Wf(j,k) = \int_{-\infty}^{+\infty} f(t) \psi_{j,k}(t) dt \qquad (6.14)$$

where j is a scaling factor, k is a shifting factor, and

$$\psi_{j,k}(t) = 2^{\frac{j}{2}} \psi(2^j t - k) \qquad (6.15)$$

are the wavelet basis functions obtained by shifting and stretching a mother wavelet $\Psi(t)$. The signal $f(t)$ can be constructed as

$$f(t) = \sum_{j=-\infty}^{+\infty} \sum_{k=-\infty}^{+\infty} (Wf(j,k)) \psi_{j,k}(t) \qquad (6.16)$$

During encoding, the wavelet transform coefficient $Wf(j, k)$ is used to replace the signal $f(t)$. Because $Wf(j, k)$ is an infinite sequence, information loss can occur in practice. Scaling functions are introduced to simplify Equation 6.16 to only contain a finite number of terms.

The main idea of applying wavelet transform to compress data is explained without providing rigorous proofs of the main results. Take the Harr wavelet as an example, and define the father of the Harr wavelet as

$$\varphi_H(t) = \begin{cases} 1, & 0 \leq t < 1 \\ 0, & \text{else} \end{cases} \tag{6.17}$$

The Harr scaling functions are obtained by shifting and stretching the father wavelet

$$\varphi_{j,k}(t) = 2^{\frac{j}{2}} \varphi_H(2^j t - k) \tag{6.18}$$

where j is an integer, and $k = 0, 1, \ldots, 2^j - 1$.

By using the definition of the Harr mother wavelet given in the example in Chapter 2, Section 2.5, that is,

$$\psi_H(t) = \begin{cases} 1, & 0 \leq t < \frac{1}{2} \\ -1, & \frac{1}{2} \leq t < 1 \\ 0, & \text{else} \end{cases} \tag{6.19}$$

it is possible to obtain the Harr wavelet functions as

$$\psi_{j,k}(t) = 2^{\frac{j}{2}} \psi_H(2^j t - k) \tag{6.20}$$

Scaling functions and wavelet functions can be shown to satisfy the following two-scale relations [10,11]:

$$\varphi_{j,0}(t) = \sum_k h_{j+1} \varphi_{j+1,k}(t) \tag{6.21}$$

$$\psi_{j,0}(t) = \sum_k g_{j+1} \varphi_{j+1,k}(t) \tag{6.22}$$

A one-dimensional signal $f(t)$ can be decomposed into the weighted combination of scaling functions in the scale j, that is,

$$f(t) = \sum_k \lambda_j(k) \varphi_{j,k}(t) \qquad (6.23)$$

Furthermore, $f(t)$ can be decomposed into the weighted combination of scaling functions and wavelet functions in the scale $j - 1$, that is,

$$f(t) = \sum_k \lambda_{j-1}(k) \varphi_{j-1,k}(t) + \sum_k \mu_{j-1}(k) \psi_{j-1,k}(t) \qquad (6.24)$$

where

$$\begin{aligned}\lambda_{j-1}(k) &= <f(t), \varphi_{j-1,k}(t)> \\ \mu_{j-1}(k) &= <f(t), \psi_{j-1,k}(t)>\end{aligned} \qquad (6.25)$$

Here, $<\bullet, \bullet>$ denotes the inner product, and

$$<f(t), \varphi_{j-1,k}(t)> = \int_{-\infty}^{+\infty} f(t) \varphi_{j-1,k}(t) dt \cdot$$

Under the same scale, scaling functions are orthogonal to each other and to the wavelet functions. Hence, each of the preceding decompositions exists and is unique. In other words, signal $f(t)$ can be uniquely fixed by using the scaling coefficients

$$\lambda_{j-1}(0), \lambda_{j-1}(1), ..., \lambda_{j-1}(2^{j-1} - 1) \qquad (6.26)$$

and the wavelet coefficients

$$\mu_{j-1}(0), \mu_{j-1}(1), ..., \mu_{j-1}(2^{j-1} - 1) \qquad (6.27)$$

Scaling coefficients represent the contour part of the signal, and wavelet coefficients represent the detailed part of the signal. In the case of Harr wavelet, calculating the coefficients of a discrete signal is an easy task.
Example 6.2: Suppose a signal $f(t)$ in the interval [0, 1] is sampled as

$$(f(0), f(1), ..., f(7)) = (15, 20, 25, 25, 30, 35, 30, 25)$$

and $f(t) = 0$ outside the interval. Compute the Harr scaling coefficients and wavelet coefficients.
Solution: $f(t)$ has 8 components in the support [0, 1]. In other words, it can be decomposed as the sum of scaling functions in the scale $j = 3$, that is,

$$f(t) = \sum_k \lambda_3(k)\varphi_{3,k}(t)$$

$$= \frac{1}{\sqrt{2^3}}[f(0)\varphi_{3,0}(t) + f(1)\varphi_{3,1}(t) + f(2)\varphi_{3,2}(t) + f(3)\varphi_{3,3}(t)$$

$$+ f(4)\varphi_{3,4}(t) + f(5)\varphi_{3,5}(t) + f(6)\varphi_{3,6}(t) + f(7)\varphi_{3,7}(t)]$$

Furthermore, the decomposition in the scale $j = 2$ is given by

$$f(t) = \sum_k \lambda_2(k)\varphi_{2,k}(t) + \sum_k \mu_2(k)\psi_{2,k}(t)$$

where

$$\lambda_2(0) = <f(t), \varphi_{2,0}(t)>$$

$$= \frac{1}{\sqrt{2^3}}[f(0)<\varphi_{3,0}(t), \varphi_{2,0}(t)> + f(1)<\varphi_{3,1}(t), \varphi_{2,0}(t)>]$$

$$= \frac{1}{\sqrt{2^3}}\left[\frac{1}{\sqrt{2}}[f(0) + f(1)]\right] = \frac{35}{\sqrt{2^4}}$$

$$\mu_2(0) = <f(t), \psi_{2,0}(t)>$$

$$= f(0)<\varphi_{3,0}(t), \psi_{2,0}(t)> - f(1)<\varphi_{3,1}(t), \psi_{2,0}(t)>$$

$$= \frac{1}{\sqrt{2^3}}\left[\frac{1}{\sqrt{2}}[f(0) - f(1)]\right] = \frac{-5}{\sqrt{2^4}}$$

Similarly,

$$\lambda_2(1) = \frac{1}{\sqrt{2^3}}\left[\frac{1}{\sqrt{2}}[f(2) + f(3)]\right] = \frac{50}{\sqrt{2^4}}$$

$$\mu_2(1) = \frac{1}{\sqrt{2^3}}\left[\frac{1}{\sqrt{2}}[f(2) - f(3)]\right] = 0$$

$$\lambda_2(2) = \frac{1}{\sqrt{2^3}}\left[\frac{1}{\sqrt{2}}[f(4) + f(5)]\right] = \frac{65}{\sqrt{2^4}}$$

$$\mu_2(2) = \frac{1}{\sqrt{2^3}}\left[\frac{1}{\sqrt{2}}[f(4) - f(5)]\right] = \frac{-5}{\sqrt{2^4}}$$

$$\lambda_2(3) = \frac{1}{\sqrt{2^3}}\left[\frac{1}{\sqrt{2}}[f(6) + f(7)]\right] = \frac{55}{\sqrt{2^4}}$$

$$\mu_2(3) = \frac{1}{\sqrt{2^3}}\left[\frac{1}{\sqrt{2}}[f(6) - f(7)]\right] = \frac{5}{\sqrt{2^4}}$$

Hence, the signal $f(t)$ is reconstructed using the scaling and wavelet functions as follows:

$$f(t) = \frac{35}{\sqrt{2^4}}\varphi_{2,0}(t) + \frac{50}{\sqrt{2^4}}\varphi_{2,1}(t) + \frac{65}{\sqrt{2^4}}\varphi_{2,2}(t) + \frac{55}{\sqrt{2^4}}\varphi_{2,3}(t)$$

$$+ \frac{-5}{\sqrt{2^4}}\psi_{2,0}(t) + \frac{0}{\sqrt{2^4}}\psi_{2,1}(t) + \frac{-5}{\sqrt{2^4}}\psi_{2,2}(t) + \frac{5}{\sqrt{2^4}}\psi_{2,3}(t)$$

In practice, the scaling functions and wavelet functions of the Harr wavelet may be defined as follows:

$$\varphi_{j,k}(t) = \varphi_H(2^j t - k)$$

$$\psi_{j,k}(t) = \psi_H(2^j t - k)$$

The scaling coefficients and the wavelet coefficients in the scale $j = 2$ are given as follows:

$$\lambda_2(0) = \frac{1}{2}[f(0) + f(1)], \mu_2(0) = \frac{1}{2}[f(0) - f(1)]$$

$$\lambda_2(1) = \frac{1}{2}[f(2) + f(3)], \mu_2(1) = \frac{1}{2}[f(2) - f(3)]$$

$$\lambda_2(2) = \frac{1}{2}[f(4) + f(5)], \mu_2(2) = \frac{1}{2}[f(4) - f(5)]$$

$$\lambda_2(3) = \frac{1}{2}[f(6) + f(7)], \mu_2(3) = \frac{1}{2}[f(6) - f(7)] \quad \blacksquare$$

The scaling terms of $f(t) = \sum_k \lambda_2(k)\varphi_{2,k}(t) + \sum_k \mu_2(k)\psi_{2,k}(t)$ can be decomposed in the scale $j = 1$:

$$f(t) = \sum_m \lambda_1(m)\varphi_{1,k}(t) + \sum_m \mu_1(m)\psi_{1,k}(t) + \sum_k \mu_2(k)\psi_{2,k}(t)$$

where

$$\lambda_1(0) = \frac{1}{2}[\lambda_2(0) + \lambda_2(1)], \mu_1(0) = \frac{1}{2}[\lambda_2(0) - \lambda_2(1)]$$

$$\lambda_1(1) = \frac{1}{2}[\lambda_2(2) + \lambda_2(3)], \mu_1(1) = \frac{1}{2}[\lambda_2(2) - \lambda_2(3)]$$

The decomposition may be carried out until the scale $j = 0$.

The following algorithm summarises the Harr wavelet encoding for one-dimensional signals.

Algorithm 6.3: The Harr wavelet encoding for one-dimensional signals
Given the one-dimensional array $f(k), k = 0,1,...,n = 2^{j0} - 1$:-
    For $j = j0 - 1$ to 0
    For $k = 0$ to $2^j - 1$

$$c(k) = \frac{1}{2}(f(2k) + f(2k+1))$$

$$c(2^j + k) = \frac{1}{2}(f(2k) - f(2k+1))$$

    End-for (k)
    Copy array $c(i), i = 0,...,2^{j+1} - 1$ to $f(t), t = 0,1,...,2^{j+1} - 1$
End-for (j)
End-Algorithm

Two-dimensional images can be treated by first decomposing each row of the given image, followed by decomposing each column using Algorithm 6.3.

6.3.4 Fractal Compression

Fractal image compression is yet another kind of transform encoding. The word "fractal" is used by Mandelbrot to express the self-similarity property of an object, that is, a given geometric shape "can be subdivided in parts such that each part is a reduced-size copy of the whole" [12]. The fractal theory was first used to simulate natural scenes in the computer graphics field, in which self-similarity is described by an iterated functional system (IFS) first proposed by Hutchinson in 1981 [13]. Later, IFS was applied in image compression. The idea is to find an IFS for a given image whose fixed point is the given image. However, using a single IFS to represent the whole image is a difficult task. An alternative method is to partition the given image into nonoverlapping blocks, and find an IFS for each block [14].

As mentioned in Section 6.3.2, an image $f(i,j), 0 \le i, j \le 2^N - 1$ is partitioned into nonoverlapping blocks, that is, the intensity matrix of the image P_f defined by Equation 6.4 is partitioned into submatrices, $R_{s,t}, 0 \le s, t \le 2^{N-n}$, defined by Equation 6.7, known as *range blocks*, each of size $2^n \times 2^n$.

Suppose each range block is associated with a set of larger submatrices, $\tilde{D}_k, k = 1, \cdots, n_D$, known as *domain blocks* of f, and are usually chosen to be of size $2^{n+1} \times 2^{n+1}$. Simple neighbouring operations, say A, may be applied to the submatrix $\tilde{\mathbf{D}}_k$ by averaging the intensities of pairwise disjoint groups of neighbouring pixel intensities. This leads to a $2^n \times 2^n$ matrix denoted symbolically as $\mathbf{D}_k = A\tilde{\mathbf{D}}_k$, which is also known as a *codebook block*.

The concepts of range blocks and domain blocks are depicted in Figure 6.5. The submatrices $\mathbf{R}_{s,t}$ and $\mathbf{D}_k$ are collocated using a rowwise data structure, which leads to the range intensity vectors $R_{s,t}$ and the codebook intensity vectors D_k. Put the concept into a minimisation problem:

For each $\mathbf{R}_{s,t}$, find an approximate codebook block $\mathbf{D}^*$ that satisfies

$$E(R_{s,t}, D^*) = \min_k \min_{\alpha,\beta} \| R_{s,t} - (\alpha D_k + \beta I) \|_2 \qquad (6.28)$$

where $R_{s,t}$ is the intensity vector with respect to the range block $\mathbf{R}_{s,t}$, and D^* is the intensity vector with respect to the codebook block $\mathbf{D}^*$.

Suppose $m = 2^n$, $R_{s,t} = (r_1, \ldots, r_m)$, and $D_k = (d_{k1}, d_{k2}, \ldots, d_{km})$, it is possible to derive the relations for scaling factor α and offset β as follows:

$$\alpha = \begin{cases} \dfrac{m \sum_{i=1}^{m}(d_{ki} \times r_i) - \left(\sum_{i=1}^{m} d_{ki}\right)\left(\sum_{i=1}^{m} r_i\right)}{m \sum_{i=1}^{m} d_{ki}^2 - \left(\sum_{i=1}^{m} d_{ki}\right)^2} & \text{if } m \sum_{i=1}^{m} d_{ki}^2 - \left(\sum_{i=1}^{m} d_{ki}\right)^2 \neq 0 \\ \\ 0 & \text{if } m \sum_{i=1}^{m} d_{ki}^2 - \left(\sum_{i=1}^{m} d_{ki}\right)^2 = 0 \end{cases} \qquad (6.29a)$$

$$\beta = \frac{1}{m}\left(\sum_{i=1}^{m} r_i - \alpha \sum_{i=1}^{m} d_{ki}\right) \qquad (6.29b)$$

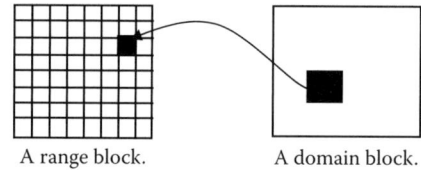

A range block.   A domain block.

FIGURE 6.5 The image f is partitioned into range blocks and domain blocks.

The *rms* error, $E(R_{s,t}, D_k)$, between $\alpha D_k + \beta I$ and $R_{s,t}$ is given by

$$E(R_{s,t}, D_k) = \sqrt{\frac{1}{m}\left[\sum_{i=1}^{m} r_i^2 + \alpha\left(\alpha \sum_{i=1}^{m} d_{ki}^2 - 2\sum_{i=1}^{m} d_{ki} r_i + 2\beta \sum_{i=1}^{m} d_{ki}\right) + \beta\left(m\beta - 2\sum_{i=1}^{m} d_{ki}\right)\right]}$$

(6.30)

Algorithm 6.4 presents the fundamental steps in a fractal image compression [15] based on the use of a fixed-size partition. The symbol $\aleph$ represents the set of all range vectors with respect to range blocks, $\aleph = \{X_{s,t}; 0 \le s, t \le 2^{N-n} - 1\}$, and Φ denotes the set of all codebook vectors with respect to codebook blocks.

Algorithm 6.4: A fractal compression method based on a fixed-size partition

Given the image $f(i,j), 0 \le i, j \le 2^N - 1$:-

Prepare $\aleph$ and Φ;

For each $R \in \aleph$ do

    For each $D_k \in \Phi$ do

        $(\alpha_k, \beta_k) :=$ Solve $\min_{\alpha, \beta} \| R - (\alpha D_k + \beta) \|_2$;

        Compute $E(R, D_k)$ using equation (6.30);

    End-For

    Compute the compression code:

        $(\alpha_{opt}, \beta_{opt}) := \min_{(\alpha_k, \beta_k)} \{E(R, D_k)\}$

End-For

End-Algorithm

When the size of the partition is fixed, that is, the size of all range blocks is the same, the algorithm does not adequately reduce spatial redundancy in images. In practice, the rate of pixel intensity variation is not maintained constant. The compression qualities may not be changed, but the compression ratio may be improved if the size of partition varies in different regions in a given image. The following adaptive fractal image compressing method using quadtree partition encapsulates the concept of adaptive partition compression [9,14]. In the algorithm, p_max is used to denote the maximal partition, which means the size of the range blocks is the biggest among all the partitions. Similarly, p_min is used to denote the minimal partition, which means the size of range blocks produced in this partition is smallest among all the partitions. Symbols $\aleph^p$ and Φ^p are used to denote the set of all range vectors and the set of all codebook vectors, respectively, according to the given partition p.

Algorithm 6.5: Adaptive fractal image compression using quadtree partition
Given the image $f(i,j), 0 \leq i, j \leq 2^N - 1$:-
Prepare the tolerance ε, the maximal partition ρ_max, and the minimal partition ρ_min;
For every possible partition ρ, prepare Φ^ρ;
For each $R \in \aleph^{\rho\_max}$ do
    $\rho = \rho$_max; $R^\rho = R$;
    Call Quadtree (ρ, R^ρ)
End-For
End-Algorithm
Procedure Quadtree (ρ, R^ρ):
$e^\rho = 10000$;
While $(e^\rho > \varepsilon)$ and $(\rho \neq \rho$_min$)$ do
    For each $D_k \in \Phi^\rho$ do
        $(\alpha, \beta) :=$ Solve $\min_{\alpha,\beta} \| R^\rho - (\alpha D_k + \beta \mathbf{I}) \|$;
        Compute $E(R^\rho, D_k)$;
    End-For;
    Compute the minimal **rms** error:-
        $e^\rho := E(R^\rho, D_{opt}) = \min\{E(R^\rho, D_k) | D_k \in \Phi^\rho\}$;
    If $(e^\rho \leq \varepsilon)$ or $(\rho = \rho$_min$)$ then
        Store tag bit 0;
        Store $\alpha_{opt}, \beta_{opt}$ and the index of D_{opt};
    Else
        Store tag bit 1;
        New Partition $\tilde{\rho}$:- Partition R^ρ into 4 quadrants;
        For each quadrant $\tilde{\mathbf{R}}$:- Call Quadtree $((\tilde{\rho}, \tilde{\mathbf{R}})$;
    End-If
End-While
End-Procedure

6.4 IMAGE COMPRESSION STANDARDS: JPEG AND MPEG

The two abbreviations—JPEG and MPEG—cannot be avoided in image compression. JPEG is the acronym for the Joint Photographic Experts Group [16]. The image compression standard coformulated by the JPEG committee and ITU-T (the predecessor of International Telegraph and Telephone Consultative Committee [CCITT]) also adopts the name JPEG. JPEG standard defines the image format file having the suffix .jpg. MPEG is the abbreviation of Moving Pictures Experts Group, which is a working group of ISO/IEC. The group aims to develop standards for video and audio compression.

In addition to JPEG and MPEG, the following image compression standards are also well known:

1. JBIG standard [16]: Designed by Joint Bi-level Image Experts Group for binary image compression.

2. H.26X standards [17]: A family of video coding standards, including H.261, H.262, H.263, and H.264. These standards, developed by ITU-T Video Coding Experts Group, are designed for transmission over ISDN lines on which data rates are multiples of 64 kbps.

6.4.1 The JPEG Standard

One common JPEG standard is ISO/IEC IS 10918-1| ITU-T Recommendation T.81 [16,18,19]. The standard includes two basic compression methods: the DCT-based method and the predictive method.

6.4.1.1 DCT-Based Method

In DCT-based compression, all input grey-scale images are partitioned into nonoverlapped blocks each of size 8×8. A stream of 8×8 blocks of grey-scale images is the input to the encoding system, which includes forward DCT (FDCT), quantisation, and entropy encoding, and Huffman encoding or arithmetical encoding may be used. Figure 6.6 depicts the encoding system. Note that for colour images, DCT-based compression is used in every single component of the image.

6.4.1.2 Predictive Method

In practice, DCT-based compression is a lossy compression method because of the quantisation process. JPEG uses a simple predictive method to meet the requirements of lossless compression. For a given grey-scale image f,

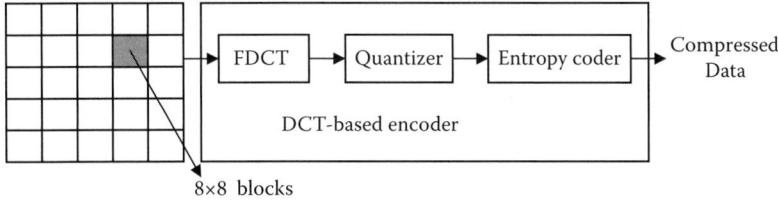

FIGURE 6.6 DCT-based encoding system.

(i − 1, j − 1)	(i − 1, j)
(i, j − 1)	(i, j)

FIGURE 6.7   The related neighbourhood of the reference pixel at.

the prediction $\tilde{f}(i, j)$ of the pixel intensity $f(i, j)$ may be formed by combing the intensities of neighbouring pixels at positions $(i − 1, j − 1)$, $(i − 1, j)$, and $(i − 1, j)$. Figure 6.7 shows the typical neighbouring pixels of the reference pixel at (i, j). The difference between $f(i, j)$ and $\tilde{f}(i, j)$ is encoded by an entropy compression method (Huffman method or arithmetic method), which produces the corresponding compressed data. Figure 6.8 depicts the predictive compression system in JPEG.

6.4.2   The MPEG Standard

There is a series of MPEG standards: MPEG-1, MPEG-2, MPEG-4, MPEG-7, and MPEG-21. MPEG-1 is the initial video and audio compression standard, and the others are extensions of MPEG-1. The basic MPEG standard includes video compression and audio compression. There are two encoding methods used for video compression: intraframe encoding and interframe encoding [21]. The former is used to reduce spatial redundancy, and the latter is used to reduce temporal redundancy. If the current frame is similar to the previous frame, interframe encoding is used; otherwise, intraframe encoding is used.

In the MPEG standard, the frames of a video sequence are partitioned into three classes: *I-frames* (intraframe), *P-frames* (prediction frame), and *B-frames* (bidirectional interpolated prediction frame).

If the current processing frame is the first frame of the video sequence or quite different from the previous frame, it is called an I-frame and is

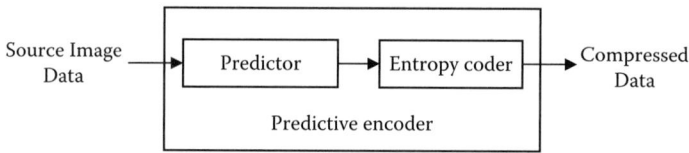

FIGURE 6.8   Predictive encoding system.

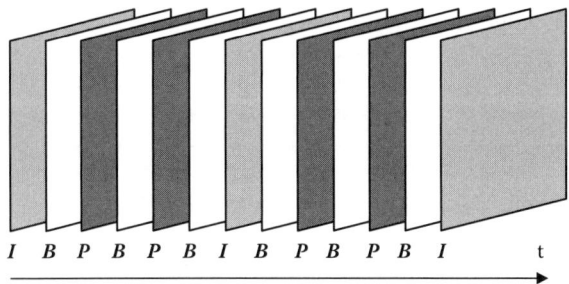

FIGURE 6.9 Three types of frames in the MPEG standard.

encoded by using intraframe encoding, which is similar to JPEG still image compression methods. If the current processing frame is similar to the previous frame, it is indexed as either a P-frame or a B-frame and is encoded by using interframe encoding. A P-frame encoding uses the previous I-frame or P-frame as a reference frame. The current P-frame being processed is predicted from the reference frame by using motion estimation. The difference between the current frame and its prediction is encoded by using DCT-based compression. It is not necessary to encode a B-frame in the source compressing system; it is reconstructed by interpolation using the previous I-frame or P-frame and the successive I-frame or P-frame in the end decompression system. The use of B-frames can improve the compression ratio effectively. Figure 6.9 depicts the three types of frames in a video sequence.

6.5 FURTHER READING

Image compression is a process requiring the use of certain technologies relatively independent from other technologies involved in image processing. Shannon's theorem in information theory provides the limit of a lossless encoding [22,23]. Among the lossy compression methods, the prevalent wavelet and fractal compression methods possess very high compression rates. Wavelet image compression possesses a multiresolution nature, which is advantageous to progressive transmission. The new standard, JPEG 2000, defined by the JPEG Committee adopts wavelet methods [24]. See References 25 and 26 for more wavelet image compression technologies. In recent years, fractal image compression technology has extended its use in fractal video compression [14]. However, fractal compression is nonsymmetric, that is, the compression process has a high computational complexity, whereas the decompression process is simple and fast. The huge amount of calculation in the compression process deters its usage in industrial applications.

6.6 EXERCISES

Q.1 Construct the Huffman codewords for the source symbols occurring in the following sentence:

This is a textbook for image processing.

Q.2 A block of the original image is given by the intensity matrix f. Its processed version is given by another intensity matrix g. Compare the PSNR values of the two versions of image blocks.

$$f = \begin{bmatrix} 206 & 112 & 24 & 18 & 50 & 70 & 17 & 45 \\ 192 & 162 & 61 & 50 & 57 & 40 & 33 & 28 \\ 164 & 173 & 176 & 39 & 66 & 11 & 25 & 27 \\ 174 & 149 & 164 & 124 & 60 & 53 & 38 & 38 \\ 192 & 164 & 159 & 207 & 107 & 45 & 46 & 76 \\ 177 & 163 & 161 & 212 & 189 & 73 & 24 & 81 \\ 109 & 170 & 151 & 186 & 186 & 117 & 19 & 61 \\ 120 & 177 & 172 & 153 & 157 & 199 & 83 & 26 \end{bmatrix}$$

$$g = \begin{bmatrix} 204 & 112 & 22 & 19 & 53 & 69 & 18 & 45 \\ 190 & 163 & 59 & 51 & 54 & 39 & 30 & 31 \\ 168 & 170 & 173 & 40 & 62 & 12 & 27 & 28 \\ 173 & 152 & 163 & 124 & 59 & 51 & 39 & 37 \\ 189 & 161 & 161 & 208 & 106 & 46 & 43 & 76 \\ 176 & 165 & 159 & 210 & 188 & 72 & 23 & 78 \\ 110 & 169 & 154 & 186 & 186 & 116 & 20 & 60 \\ 119 & 174 & 171 & 155 & 156 & 199 & 82 & 26 \end{bmatrix}$$

Q.3 By applying the one-dimensional wavelet image compression described in Algorithm 6.3 twice, first along the rowwise direction followed by the columnwise direction, to the image block defined by the intensity matrix f in Q.2, obtain the resulting intensity matrix.

Q.4 Partition the image block as defined by the intensity matrix f in Q.2 into range blocks each of size 4×4. Take the entire 8×8 matrix as the domain block. Compute the scaling α and the offset β between the top left range block and the domain block by using Equations 6.29a, 6.29b.

6.7 REFERENCES

1. A. N. Netravali and B. G. Haskell, *Digital Pictures: Representation, Compression and Standards*, 2nd edition, Plenum Press, New York, 1995.
2. T. H. Cormen, C. E. Leiserson, L. R. Ronald, et al., *Introduction to Algorithms*, 2nd Edition, MIT Press and McGraw-Hill, 2001.
3. K. Sayood, *Introduction to Data Compression*. 2nd edition, Morgan Kaufmann, 2000.
4. R. Qiuqi, *Digital Image Processing Science*, Publishing House of Electronic Industry, Beijing, 2001 (in Chinese).
5. W. N. Waggener, *Pulse Code Modulation Systems Design*, Boston, MA, Artech House, 1999.
6. M. Wang and C.-H. Lai, A hybrid fractal video compression method, *Computers and Mathematics with Applications*, Vol. 50(3–4): 611–621, Elsevier Science, 2005.
7. A. Gersho and R. M. Gray, *Vector Quantization and Signal Compression*, Springer, 1991.
8. http://www.data-compression.com/vq.shtml#intro.
9. D. Saupe, R. Hamzaoui, and H. Hartenstein, Fractal image compression—an introductory overview, *In Fractal Models for Image Synthesis, Compression and Analysis, ACM SIGGRAPH'96 Course Notes 27* (Ed: D. Saupe and J. Hart), New Orleans, Louisiana, August 1996.
10. C. S. Burrus and R. A. Gopinath, *Introduction to Wavelets and Wavelet Transforms*, Prentice Hall, 1997.
11. Y. Sheng, Wavelet Transform, In *The Transforms and Applications Handbook*, Ed. by A. D. Poularikas, pp. 747–827, CRC Press, 1996.
12. B. B. Mandelbrot, *The Fractal Geometry of Nature,* W. H. Freeman, 1982.
13. J. E. Hutchinson, Fractal and self-similarity, *Indiana University Mathematics Journal*, Vol. 30, pp. 713–747, September–October 1981.
14. M. Wang and C. H. Lai, Gray video compression methods using fractals, *International Journal of Computer Mathematics*, Vol. 84(11): 1567–1590, November 2007.
15. M. Wang, C.-H. Lai, A hybrid fractal video compression method, *Computers and Mathematics with Applications*. Volume 50, Issues 3–4: 611–621, Elsevier Science, 2005.
16. http://www.jpeg.org/.
17. http://www.itu.int/rec/T-REC-H.263/.

18. G. K. Wallace, The JPEG still compression standard, *Communications of ACM*, Vol. 34(4): 30–44, 1991.
19. Digital Compression and Coding of Continuous-tone Still Images, Part 1, Requirements and Guidelines. ISO/IEC JTC1 Draft International Standard 10918-1, November 1991.
20. R. Steinmetz and K. Nahrstedt, *Multimedia: Computing, Communications and Applications,* Prentice Hall PTR, 1995.
21. C. Fogg, D. J. LeGall, J. L. Mitchell, and W. B. Pennebaker, *MPEG Video Compression Standard*, Spring, 1996.
22. C. E. Shannon, A mathematical theory of communication, *Bell System Tech. Journal*, 27: 379–423, 623–656, 1948.
23. C. E. Shannon, A. D. Wyner, and N. J. A. Sloane, *Clause E. Shannon: Collected Papers*, Wiley—IEEE Press, New York, 1993.
24. C. Christopoulos, A. Skodras, and T. Ebrahimi, The JPEG2000 still image coding system: an overview, *IEEE Transaction on Consumer Electronics*, Vol. 46(4): 1103–1127, 2000.
25. J. M. Shapiro, Embedded image coding using zerotree of wavelet coefficients, *IEEE Transaction on Signal Processing*, Vol. 41(12): 3445–3462, 1993.
26. A. Said and W. A. Pearlman, A new, fast, and efficient image codec based on set partitioning in hieratchical trees, *IEEE Transaction on Circuits Syst. Video Technology*, Vol. 6(3):243–250, 1996.

6.8   PARTIAL CODE EXAMPLES

Project 6-1: Huffman Encoding

(These codes can be found in CD: Project6-1\source code\Project6-1View .cpp and DlgHuffman.cpp)

```
#include "stdafx.h"
#include "project6_1.h"
#include "DlgHuffman.h"
#include "project6_1Doc.h"
#include "project6_1View.h"
#ifdef _DEBUG
#define new DEBUG_NEW
#undef THIS_FILE
static char THIS_FILE[] = __FILE__;
#endif
/************************************************************
********
*  Function name:
*  OnHuffmanEncoding()
*
*  Parameter:
```

```
 *	None
 *
 *	Return Value:
 *	None
 *
 *	Description:
 *	Huffman encoding
 *
 ***************************************************************
 ******/
void Cproject6_1View::OnHuffmanEncoding()
{
        // Get the document
        CProject7_1Doc* pDoc = GetDocument();

        // the pointer pointing to DIB's pixel
        LPSTR  lpDIB;

        // the pointer pointing to the DIB's pixel
        LPSTR  lpDIBBits;

        // Lock DIB
        lpDIB = (LPSTR) ::GlobalLock((HGLOBAL)
pDoc->GetHDIB());

 // Find the outset position of the DIB's image pixel
        lpDIBBits = ::FindDIBBits(lpDIB);
        int cxDIB = (int) ::DIBWidth(lpDIB); // Size of
DIB - x
        int cyDIB = (int) ::DIBHeight(lpDIB); // Size of
DIB - y
        long lLineBytes = WIDTHBYTES(cxDIB * 8);
        // count the number of bytes of the image per line

        // Change the shape of the cursor
        BeginWaitCursor();
        // the pointer pointing to the original image
        unsigned char *    lpSrc;

        // the width and the height of the image
        LONG   lHeight;
        LONG   lWidth;
```

```
        // total pixel number of the image
        LONG    lCountSum;

        // cyclic variables
        LONG    i;
        LONG    j;

        // array used for saving the probabilities of each
grey level
        double * dProba;

        // the colour number of the current image
        int         nColourNum;

        // change the style of the cursor
        BeginWaitCursor();


/*******************************************************
         * Compute the probabilities of grey levels occur-
ring in the image
         *          ********************************************
**************************
         */

        // get the store bits per pixel used for colour
information from the head file
        nColourNum = ::DIBNumColours(lpDIB);
        // allocate memory
        dProba = new double[nColourNum];

        //width and height of the image
lWidth = cxDIB;
        lHeight = cyDIB;
        // total pixel number
        lCountSum = lHeight * lWidth;

        // assign each probability variable to 0
        for (i = 0; i < nColourNum; i ++)
        {
                dProba[i] = 0.0;
        }
```

```
        // count the occurring number of each grey level
        for (i = 0; i < lHeight; i ++)
        {
            for (j = 0; j < lWidth; j ++)
            {
                // the pointer pointing to the i-th
line and j-th picture element
                lpSrc = (unsigned char*)lpDIBBits +
lLineBytes * (cyDIB - 1 - i) + j;

                // add a to the count
                dProba[*(lpSrc)] = dProba[*(lpSrc)] +
1;
            }
        }


        // compute the occurring probability of each scale
level in the image
        for (i = 0; i < nColourNum; i ++)
        {
            dProba[i] = dProba[i] / (FLOAT)lCountSum;
        }

        /***********************************************
         * construct the Huffman codeword table and show it
in a dialogue box
         ***********************************************/

        // construct the dialogue box
        CDlgHuffman dlgCoding;

        // initialise the variable
        dlgCoding.dProba = dProba;
        dlgCoding.nColourNum = nColourNum;

        // show the dialogue box
        dlgCoding.DoModal();

        // restore the style of the cursor
        EndWaitCursor();
}
```

```cpp
/***********************************************************
***
* DlgHuffman.cpp : implementation file
***********************************************************
**/
#include "stdafx.h"
#include "project6_1.h"
#include "DlgHuffman.h"
#include <math.h>
#ifdef _DEBUG
#define new DEBUG_NEW
#undef THIS_FILE
static char THIS_FILE[] = __FILE__;
#endif
/////////////////////////////////////////////////////////////
///////////////////////
// CDlgHuffman dialogue
CDlgHuffman::CDlgHuffman(CWnd* pParent /*=NULL*/)
       : CDialogue(CDlgHuffman::IDD, pParent)
{
       //{{AFX_DATA_INIT(CDlgHuffman)
       m_dEntropy = 0.0;
       m_dCodLength = 0.0;
       m_dRatio = 0.0;
       //}}AFX_DATA_INIT
}
void CDlgHuffman::DoDataExchange(CDataExchange* pDX)
{
       CDialogue::DoDataExchange(pDX);
       //{{AFX_DATA_MAP(CDlgHuffman)
       DDX_Control(pDX, IDC_LIST2, m_lstTable);
       DDX_Text(pDX, IDC_EDIT1, m_dEntropy);
       DDX_Text(pDX, IDC_EDIT2, m_dCodLength);
       DDX_Text(pDX, IDC_EDIT3, m_dRatio);
       //}}AFX_DATA_MAP
}
BEGIN_MESSAGE_MAP(CDlgHuffman, CDialogue)
       //{{AFX_MSG_MAP(CDlgHuffman)
       //}}AFX_MSG_MAP
END_MESSAGE_MAP()
/////////////////////////////////////////////////////////////
///////////////////////
```

```
// CDlgHuffman message handlers
BOOL CDlgHuffman::OnInitDialog()
{
        // default OnInitDialogue()
        CDialog::OnInitDialogue();
        // cyclic variables
        LONG    i;
        LONG    j;
        LONG    k;

        // temporary variable
        double dT;

        // string variable
        CString     str2View;

        // the item of the Control ListCtrl
        LV_ITEM lvItem;

        // used for saving the new item number of the
control ListCtrl
        int             nItem2View;
        // array used for saving temporary results
        double *    dTemp;

        // array used for saving the mapping between the
grey level and the position
        int     *       n4Turn;

        // initialise the variables
        m_dEntropy = 0.0;
        m_dCodLength = 0.0;

        // allocate the memory
        m_strCode = new CString[nColourNum];
        n4Turn = new int[nColourNum];
        dTemp = new double[nColourNum];

        // assign values to dTemp
        // arrange the grey level in ascending order
        for (i = 0; i < nColourNum; i ++)
        {
```

```
                dTemp[i] = dProba[i];
                n4Turn[i] = i;
        }
        // sort the probabilities of grey levels using the bubble sort method
        // and change the position in the mapping
        for (j = 0; j < nColourNum - 1; j ++)
        {
                for (i = 0; i < nColourNum - j - 1; i ++)
                {
                        if (dTemp[i] > dTemp[i + 1])
                        {
                                dT = dTemp[i];
                                dTemp[i] = dTemp[i + 1];
                                dTemp[i + 1] = dT;

                                // swap the position of the grey level i with the one of i+1
                                for (k = 0; k < nColourNum; k ++)
                                {
                                        if (n4Turn[k] == i)
                                                n4Turn[k] = i + 1;
                                        else if (n4Turn[k] == i + 1)
                                                n4Turn[k] = i;
                                }
                        }
                }
        }

        /*********************************************************
        * construct the Huffman codeword table
        **********************************************************/

        // begin from the probability > 0
        for (i = 0; i < nColourNum - 1; i ++)
        {
                if (dTemp[i] > 0)
                        break;
```

```
        }

    for (; i < nColourNum - 1; i ++)
    {
        // update m_strCode
        for (k = 0; k < nColourNum; k ++)
        {
            // check the grey level i
            if (n4Turn[k] == i)
            {
                // if the grey level is small, the codeword add "1"
                m_strCode[k] = "1" + m_strCode[k];
            }
            else if (n4Turn[k] == i + 1)
            {
                // if the grey level is bigger, the codeword add "0"
                m_strCode[k] = "0" + m_strCode[k];
            }
        }

        // save the sum of two minimum probabilities to dTemp[i + 1]
        dTemp[i + 1] += dTemp[i];

        // change the mapping
        for (k = 0; k < nColourNum; k ++)
        {
            // change the position i of the grey level i
            // as the position of the grey level i+1
            if (n4Turn[k] == i)
                n4Turn[k] = i + 1;
        }

        // rearrange
        for (j = i + 1; j < nColourNum - 1; j ++)
        {
            if (dTemp[j] > dTemp[j + 1])
```

```cpp
            {
                        // swap
                        dT = dTemp[j];
                        dTemp[j] = dTemp[j + 1];
                        dTemp[j + 1] = dT;

                        // // swap the positions of the grey level i and i+1
                        for (k = 0; k < nColourNum; k ++)
                        {
                                if (n4Turn[k] == j)
                                    n4Turn[k] = j + 1;
                                else if (n4Turn[k] == j + 1)
                                    n4Turn[k] = j;
                        }
                }
                else
                // exit the cycle
                        break;
            }
        }

        // compute the entropy of the image
        for (i = 0; i < nColourNum; i ++)
        {
            if (dProba[i] > 0)
            {
                m_dEntropy -= dProba[i] * log(dProba[i]) / log(2.0);
            }
        }
        // compute the average length of the codewords
        for (i = 0; i < nColourNum; i ++)
        {
            // sum
            m_dCodLength += dProba[i] * m_strCode[i].GetLength();
        }

        // compute the efficiency of the encoding
```

```
        m_dRatio = m_dEntropy / m_dCodLength;

        // save the change
        UpdateData(FALSE);

        /*****************************************************
        * output the result
        *****************************************************/

        // set the style of the control CListCtrl
        m_lstTable.ModifyStyle(LVS_TYPEMASK, LVS_REPORT);

        // add a header to the control List
        m_lstTable.InsertColumn(0, "Grey level", LVCFMT_
LEFT, 60, 0);
        m_lstTable.InsertColumn(1, "Probability", LVCFMT_
LEFT, 78, 0);
        m_lstTable.InsertColumn(2, "Huffman codeword",
LVCFMT_LEFT, 110, 1);
        m_lstTable.InsertColumn(3, "length of the code-
word", LVCFMT_LEFT, 78, 2);
        // set the style of the Control as text
        lvItem.mask = LVIF_TEXT;

        // add items
        for (i = 0; i < nColourNum; i ++)
        {
                //add the first item
                lvItem.iItem = m_lstTable.GetItemCount();
                str2View.Format("%u",i);
                lvItem.iSubItem = 0;
                lvItem.pszText= (LPTSTR)(LPCTSTR)str2View;
                nItem2View = m_lstTable.InsertItem(&lvItem);

                // add the other items
                lvItem.iItem = nItem2View;

                // add the probability of the grey level
                lvItem.iSubItem = 1;
                str2View.Format("%f",dProba[i]);
                lvItem.pszText = (LPTSTR)(LPCTSTR)str2View;
                m_lstTable.SetItem(&lvItem);
```

```
            // add the Huffman codeword
            lvItem.iSubItem = 2;
            lvItem.pszText = (LPTSTR)(LPCTSTR)m_
strCode[i];
            m_lstTable.SetItem(&lvItem);

            // add the length of the codeword
            lvItem.iSubItem = 3;
            str2View.Format("%u",m_strCode[i].
GetLength());
            lvItem.pszText = (LPTSTR)(LPCTSTR)str2View;
            m_lstTable.SetItem(&lvItem);
      }
      // release memory
      delete n4Turn;
      delete dTemp;
      // return TRUE
      return TRUE;
}
```

Project 6-2: Fractal Image Compression

```
(These codes can be found in CD: Project6-3\source code\
Project6-3View.cpp)
#include "stdafx.h"
#include "project6_3.h"
#include "CMP.h"
#include "math.h"
#include "project6_3Doc.h"
#include "project6_3View.h"
#include "DECMPdlg.h"
#ifdef _DEBUG
#define new DEBUG_NEW
#undef THIS_FILE
static char THIS_FILE[] = __FILE__;
#endif
/************************************************************
*********
* Function name:
* OnFractalCompress()
*
* Parameter:
```

```
 * None
 *
 * Return Value:
 * None
 *
 * Description:
 * Fractal Compress
 *
 ***********************************************************
 *******/
void CProject6_3View::OnFractalCompress()
{
        // Change the shape of the cursor
        BeginWaitCursor();
        // construct the dialogue box
        CCMP dlgCmp;

        // show the dialogue box
if (dlgCmp.DoModal() == IDOK && dlgCmp.m_FileIn!=_T("")
&& dlgCmp.m_FileOut!=_T(""))
{
        unsigned int tmp=0, tj=0, wdata=0;
        int cxDIB = 256; // Size of DIB - x
        int cyDIB = 256; // Size of DIB - y
long lLineBytes = 256;
 // count the number of bytes of the image per line
        unsigned char *lpDIBBits;
 lpDIBBits=(unsigned char *)malloc(sizeof(unsigned char)
* 256*256);
        //get the input file and the output file
        FILE *rFile=fopen(dlgCmp.m_FileIn,"rb");
        FILE *wFile=fopen(dlgCmp.m_FileOut,"wb");
        unsigned int Offset;
        fseek (rFile,10,0);
        fread (&Offset,4,1,rFile);
        fseek (rFile,Offset,0);
        fread (lpDIBBits,cxDIB*cyDIB,1,rFile);
        fclose (rFile);

        int i, j, i1, j1;
        int n=2; //the size of the block
        int m = n*n;
```

```cpp
// partition the original image
unsigned char*** R=new unsigned char**[cyDIB/n];
for (i=0; i<cyDIB/n; i++)
{
        R[i] = new unsigned char* [cxDIB/n];
        For (int j=0; j<cxDIB/n; j++)
            R[i][j]=new unsigned char[n*n];
}
for (i=0; i<cyDIB; i++)
{
        for (j = 0; j<cxDIB; j++)
        {
            R[int(i/n)][int(j/n)][(i%n)*n+j%n]= GetData((unsigned char*)lpDIBBits,j,i,lLineBytes);
        }
}
int Dlinenum = cxDIB-2*n+1;// the domain number in every line
unsigned char** D = new unsigned char* [(cxDIB-2-*n+1)*(cyDIB-2*n+1)];
for (i=0; i<(cxDIB-2*n+1)*(cyDIB-2*n+1); i++)
        D[i]=new unsigned char [n*n];
int index=0;
for (i=0; i<cyDIB-2*n; i++)
{
        for (j=0; j<cxDIB-2*n; j++)
        {
            for (i1=0; i1<n; i1++)
            {
                for (j1=0; j1<n; j1++)
                {
D[index][i1*n+j1]= GetData((unsigned char*)lpDIBBits,2*j1,2*i1,lLineBytes)+
                        GetData((unsigned char*)lpDIBBits,2*j1+1,2*i1,lLineBytes)+
                        GetData((unsigned char*)lpDIBBits,2*j1,2*i1+1,lLineBytes)+
GetData((unsigned char*) lpDIBBits, 2*j1+1, 2*i1+1, lLineBytes) ;
                        D[index][i1*n+j1]/=4;
                }
            }
```

```
                index++;
            }
        }
        //initialise the compression data array
        int **x=new int* [cyDIB/n];// the initial line
position of the domain
        int **y=new int* [cyDIB/n];// the initial column
position of the domain
        int    **ki=new int* [cyDIB/n];//
        int    **g=new int* [cyDIB/n];// error
        for(i=0;i<cyDIB/n;i++)
        {
            x[i]=new int [cxDIB/n];
            y[i]=new int [cxDIB/n];
            ki[i]=new int [cxDIB/n];
            g[i]=new int [cxDIB/n];
        }
        double rsum, dsum, rdsum, r2sum, d2sum;
        int trans_x[4]={1,1,-1,-1}; //x-parameters of
four transforms
        int trans_y[4]={1,-1,1,-1}; // y-parameters of
four transforms
        for (i=0; i<cyDIB/n; i++)
        {
            for (j=0; j<cxDIB/n; j++)
            {
                int dg=0; // the average error of the
block R and the current domain D                    dou-
ble det=9999999999999; // the error of R and D
                double alpha, beta;
                // search the best domain
                for (int i1=0; i1<index; i1++)
                {
                    if (i1>600)
                        break;
                    rsum=0; // sum of the intensi-
ties of R
                    dsum=0; //sum of the intensi-
ties of D
                    rdsum=0;
                    r2sum=0;
                    d2sum=0;
```

```
                                for (int j1=0; j1<n*n; j1++)
                                {
                                        rsum += R[i][j][j1];
                                        dsum += D[i1][j1];
                                }
                                dg = int((rsum-dsum)/(n*n));
                                unsigned char* g1=new unsigned
char [n*n];

                                for (j1=0; j1<n*n; j1++)
                                {
                                        g1[j1] = D[i1][j1]+dg;
                                        if (g1[j1] > 255)
                                                g1[j1] = 255;
                                        else if (g1[j1] < 0)
                                                g1[j1] = 0;
                                }
                                for (int t=0; t<4; t++)
                                {
                                        Rotate(g1);
                                        rsum=0;
                                        dsum=0;
                                        rdsum=0;
                                        r2sum=0;
                                        d2sum=0;

                                        for (j1=0; j1<n*n; j1++)
                                        {
                                                rsum += R[i][j]
[j1];
                                                dsum += g1[j1];
                                                rdsum += R[i][j]
[j1]*g1[j1];
                                                r2sum += R[i][j]
[j1]*R[i][j][j1];
                                                d2sum +=
g1[j1]*g1[j1];
                                        }
                                        double temp = (m*d2sum-
dsum*dsum);
                                        if (temp != 0)
                                                alpha =
(m*rdsum-dsum*rsum)/temp;
```

```
                                else
                                        alpha=0;
                                beta = (rsum-
alpha*dsum)/m;
double detsum = sqrt((r2sum + alpha* (alpha*d2sum
- 2*rdsum
+ 2*beta*dsum) + beta*(m*beta - 2*rsum))/m);
                                if (detsum < det)
                                {
                                        ki[i][j] = (t+1)%4;
                                        g[i][j] = dg;
                                        x[i][j] =
i1%Dlinenum;
                                        y[i][j] = int(i1/
Dlinenum);
                                        det = detsum;
                                }
                        }
                        delete [] g1;
                }
tmp =(x[i][j]<<19) ^ (y[i][j]<<11) ^ (((((g[i]
[j]>>31)&1)<<8) ^ (abs(g[i][j]))) <<2 ) ^ ( ki[i][j]);
                        tj += 27;
                        if (tj < 32)
                        {
                                wdata ^= tmp<<(32-tj);
                        }
                        else
                        {
                                wdata^=tmp>>(tj-32);
                                fwrite (&wdata,4,1,wFile);
                                tj -= 32;
                                wdata = (tmp&((1<<tj)-1))<<(32-
tj);
                        }
                   }
              }
        }
        free (lpDIBBits);
        fclose (wFile);
        for (i=0;i<index;i++)
                delete [] D[i];
        delete [] D;
```

```
        for(i=0;i<cyDIB/n;i++){
            for(int j=0;j<cxDIB/n;j++)
                delete [] R[i][j];
            delete [] R[i];
            delete [] x[i];
            delete [] y[i];
            delete [] ki[i];
            delete [] g[i];
        }
        delete [] R;
        delete [] x;
        delete [] y;
        delete [] ki;
        delete [] g;
    }

    // Reset the shape of the cursor
    EndWaitCursor();
}
```

Index

A

Adaptive method, 138
Adaptive morphology, 184
Amplitude, 1
Analogue signals, 1–2
 processing, 3
 storing, 3
Anisotropic diffusion models, 98–101

B

Band thresholding, 137
Barrel distortion, 95
B-frames, 222
Bilinear interpolation, 97
Bi-model, 138
Binary erosion, 186–194
Binary images, 7, 150
 dilation, 171
 erosion, 171
 Hough transform, 151
 mathematical morphology, 169
 objects, 168
 object structure, 167
 set, 169
Binary morphological operation
 applications, 176–177
 dilation operation, 171–172
 erosion operation, 173–174
 grey-scale images, 178
 hit-or-miss transformation, 175, 176–177
 opening and closing operation, 175
 skeleton method, 177
 thinning and thickening, 176
Binary scale images, 167
Binary skeleton operation
 mathematical morphology, 195–198
 results, 178

Binary thinning operation, 177
Bit mapped (BMP) format, 17–18, 201
Bivariant polynomials, 94
Blocky effect, 99
Blur images, 92
BMP. *see* Bit mapped (BMP) format
Border tracing, 149
 detection, 155
 edge-based segmentation, 148–149
 image segmentation, 161–166
Brightness, 66
Butterworth filtering
 high-pass, 87
 low-pass, 79

C

Canny method
 detector, 148
 edge detection, 143
 edge image, 143
CCD. *see* Charge-coupled device (CCD)
CCITT. *see* International Telegraph and Telephone Consultative Committee (CCITT)
Charge-coupled device (CCD), 18
Closing operations, 176
CMOS. *see* Complementary metal-oxide semiconductor (CMOS)
CMY model, 16–17
Code book block, 218
Code vectors, 208, 210
Coding method, 206
Colour images, 7–16
 channels, 9
 CMY model, 16
 distinguishing, 8
 8-bit, 25–28

243

HSI colour model, 12–16
HSI to RGB model conversion, 14–15
RGB colour model, 9
RGB to HSI model conversion, 14
YIQ colour model, 10
YUV colour model, 11
Colour mixing based, 12
Colour planes, 9
Colour television broadcast, 10
Colour triangle, 12, 15
Complementary metal-oxide semiconductor (CMOS), 18
Completeness, 47
Compression methods, 207
Computing convolution, 35
Constrained conditional restoration, 89
Continuous Fourier transform, 37–38
  one dimension, 37–38
  two dimension, 38
Continuous Gaussian function, 98
Continuous models, 103
Continuous time domain, 32
Continuous wavelet transform, 44
Contrast stretching, 69–70
  linear transform, 70
Convolution image functions, 82
Convolution kernel
  discrete image functions, 73
  image smoothing, 48
  symmetric matrix, 36
Convolution operations, 32–36, 182
Correlation operations, 30–31
Crack edges
  concept, 144
  confidence, 146
  directions, 145
Cut-off frequency, 79

Differential pulse code modulation (DPCM), 207
  decoding process, 208
  encoding process, 208
Diffusion-based models, 98–101
  heat conduction, 98
  PDE-based image-processing methods, 97
Digital images, 81
Digital signals, 3–5
  quantisation, 5
  sampling, 4
Dilation
  algorithm, 172
  binary images, 171
  example, 172
  grey-scale images, 178–181
  operation, 171–172
Discrete cosine transform (DCT), 18
  compression, 221, 223
  encoding system, 221
  image compression, 221
  image processing tools, 42, 58–60
  inverse, 43
  JPEG standards, 221
Discrete Fourier transform (DFT), 38–40
  properties, 39–40
Discrete Gaussian smoothing convolution kernel, 98
Discrete image functions, 73
Discrete signal sequences, 33
Disjoint structuring elements, 176
Distorted images, 94
Domain blocks, 217
  image, 218
DPCM. see Differential pulse code modulation (DPCM)

D

DCT. see Discrete cosine transform (DCT)
Decoding principles, 208
Decomposition of image functions, 46
Decompression process, 210
Degradation model, 89–90
  constrained conditional restoration, 90
  noise, 92
  unconstrained conditional restoration, 89
Degraded recovery images, 87
DFT. see Discrete Fourier transform (DFT)

E

Easy-to-use coding method, 206
Edge-based segmentation, 140–151, 152
  border tracing, 148–149
  edge image thresholding, 140–142
  edge relaxation, 143–147
  Hough transform, 150–151
Edge detection
  Canny method, 143
  edge-tracing method, 150

Edge enhancement, 140
Edge image
  Canny method, 143
  segmentation, 140–142
  thresholding, 140–142
Edge magnitude, 135
Edge pixels, 141
Edge relaxation, 145
  algorithm, 147
  edge-based segmentation, 143–147
  method, 144
Edge-stopping function, 99
Edge-tracing method, 150
8-bit colour image, 25–28
8-bit grey-scale image
  conversion to red channel image, 22
  histogram, 108–112
  image preprocessing techniques, 108–112
  matrix, 180
8-neighbourhood
  border tracing detection, 155
  directions of search, 148
Eight-point decimation-in-time FFT butterfly flowchart, 42
Encoding principles, 208
End vertices, 145
Energy conversation theorem, 40
Entropy compression method, 222
Erosion
  binary images, 171
  example, 184
  grey-scale image, 182
  inverse transformation, 175
  results, 173
Erosion operations
  binary dilation, 175
  binary morphological operation, 173–174
  examples, 174
  grey-scale images, 182

F

Fast Fourier transform, 41
4-bit grey-scale image
  histogram, 155
Fourier transform, 91
  continuous, 37–38
  discrete, 38–40
  fast, 41
  Gabor transform, 43
  image processing tools, 37–41, 49–57
  one dimension, 37–38
  orthogonal transformations, 48
  properties, 39–40
  spatial domain definition, 38
  two dimension, 38
Fourth-order partial differential equations image restoration
  image preprocessing techniques, 114–134
Fractal compression, 217
  fixed-size partition, 219
  image compression, 217–219, 236–242
Frequency-domain methods, 78–79
  Butterworth low-pass filtering, 79
  ideal low-pass filtering, 78
  trapezoidal low-pass filtering, 79

G

Gabor transform, 30, 44
  Fourier transform, 43
  image processing tools, 43
Gaussian convolution kernel, 75
Gaussian filtering
  image preprocessing techniques, 75
  smoothing, 75
  spatial domain methods, 73
Gaussian function, 43
Gaussian noise, 88, 99–100
Gaussian smoothing, 98
  linear filtering method, 75
Geometric rectification, 93–96
GIF. see Graphics interchange format (GIF)
Global operations, 29
Gradient directions, 141
Gradient operators, 81–85, 140
  Laplacian operator, 84
  Prewitt operator, 83
  Roberts operator, 83
  Sobel operator, 83
Graphics interchange format (GIF)
  image storage, 19
Grey levels
  grey-scale images, 6
  homogeneity principle, 152

Grey level transformation
   contrast stretching, 69–70
   histogram equalisation, 66–68
   histogram image enhancement, 66–68
Grey-scale dilation
   algorithm, 181
   convolution operation, 182
   example, 184
Grey-scale erosion algorithm, 183
Grey-scale image, 6, 7, 67, 178–198
   applications, 183–184
   conversion to red channel image, 22
   dilation operations, 178–181
   8-bit, 108–112
   erosion, 182
   erosion operations, 182
   4-bit histogram, 155
   grey levels, 6
   histogram, 108–112
   image preprocessing techniques, 108–112
   intensity matrix, 36, 48
   mathematical morphology, 167
   matrix, 180
   operations, 178–182
   processing, 6
   resolution, 6
Grey-scale morphology, 179
   definitions, 183

H

Harr mother wavelet, 213
Harr scaling coefficients, 214
Harr scaling functions, 213
Harr wavelet, 213, 216
   basis functions, 46
   encoding, 217
   function, 45
Heat conduction model
   anisotropic diffusion models, 98–101
   diffusion-based models, 98
High-pass filtering, 85–87
   Butterworth, 87
   ideal, 85–86
   trapezoidal, 87
Histograms
   adaptive method, 138
   based method, 138
   based thresholding, 137–138
   4-bit grey-scale image, 155
   equalisation, 66, 69
   grey-level- transformation, 66–68
   image enhancement, 66–68
   map, 67
   Mode method, 138
   pixel brightness, 66–68
Hit-or-miss transformation
   binary morphological operation, 175, 176–177
Homogeneity principle
   grey level, 152
   mean-grey levels, 153
Hough transform, 156
   binary image, 151
   edge-based segmentation, 150–151
   illustration, 151
HSI. *see* Hue saturation intensity (HSI)
Hue, 8
Hue saturation intensity (HSI)
   colour model, 12–16
   to RGB model conversion, 14–15
   triangle models, 13
Huffman encoding, 202, 221
   algorithm, 204
   image compression, 201–205, 226–235
Huffman method, 221
Huffman tree, 204
   binary numbers, 205
Human visual system, 10

I

Ideal high-pass filtering, 85–86
Ideal low-pass filtering, 78
Identity operator, 102
I-frames, 222
IFS. *see* Iterated functional system (IFS)
Image(s)
   blurring, 75
   contrast, 69
   degradation model, 88
   degraded recovery, 87
   edge enhancement, 140
   fidelity metrics, 200
   filtering, 78
   intensities, 84
   PSNR values, 201

resolution, 6
restoration, 87–88
sharpening, 80
signals analysis, 78
transform, 46
transmitted, 65
two-dimensional signal function, 1
Image compression, 199–242
   DCT-based method, 221
   exercises, 224
   fractal compression, 217–219
   fractal image compression, 236–242
   Huffman encoding, 201–205, 226–235
   image fidelity metrics, 200
   JPEG standards, 221
   lossless compression, 201–206
   lossy compression, 207–219
   MPEG standards, 222
   partial code examples, 226–242
   predictive compression methods, 207
   predictive method, 221
   runlength encoding, 206
   standards, 220–222
   vector quantisation, 208–211
   wavelet compression, 212–216
Image enhancement, 71, 80–87
   Butterworth high-pass filtering, 87
   gradient, 80
   gradient image, 81
   gradient operators, 81–85
   high-pass filtering, 85–87
   ideal high-pass filtering, 85–86
   Laplacian operator, 84
   Prewitt operator, 83
   Roberts operator, 83
   Sobel operator, 83
   trapezoidal high-pass filtering, 87
Image functions
   convolution, 82
   decomposition, 46
Image preprocessing techniques, 65–134
   anisotropic diffusion models, 98–101
   based on degradation model, 89–90
   8-bit grey-scale image histogram, 108–112
   Butterworth high-pass filtering, 87
   Butterworth low-pass filtering, 79

concepts and models, 71–72
constrained conditional restoration, 90
contrast stretching, 69–70
diffusion-based models, 98–101
exercises, 105–107
frequency-domain methods, 78–79
Gaussian filtering, 75
geometric rectification, 93–96
gradient, 80
gradient image, 81
gradient operators, 81–85
grey-level- transformation, 66–70
heat conduction model, 98–101
high-pass filtering, 85–87
histogram equalisation, 66–68
histogram image enhancement, 66–68
ideal high-pass filtering, 85–86
ideal low-pass filtering, 78
image degradation model, 87–88
image enhancement, 80–87
image restoration, 87–96
image smoothing, 73–79
inverse filtering, 91
limiting linear transform, 70
linear transform, 70
median filtering, 76, 112–117
neighbourhood-averaging methods, 73–74
partial code examples, 108–134
partial differential equations processing methods, 97–134
PDE image restoration, 114–134
PDE model discrete formats, 103
pixel brightness, 66–70
pixel intensity confirmation, 96
Sobel operator gradient image, 118–123
spatial-domain methods, 73–77
spatial geometric transforms, 94–95
threshold-averaging methods, 75
trapezoidal high-pass filtering, 87
trapezoidal low-pass filtering, 79
TV-based models, 102
unconstrained conditional restoration, 89
weighted median filtering, 77
Wiener filtering, 92

Image processing, 104, 212
  enhancement, 65
  PDE, 104
  restoration, 65
  smoothing, 65
  storage, 199
  techniques, 65
  transmission, 199
Image processing tools, 29–64
  completeness, 47
  continuous Fourier transform, 37–38
  continuous wavelet transform, 44
  convolution operation, 30–36
  correlation operation, 30–36
  DCT transformation, 58–60
  discrete cosine transform, 42
  discrete Fourier transform, 38–40
  discrete wavelet transform, 45
  exercises, 48
  fast Fourier transform, 41
  Fourier transform, 37–41, 49–57
  gabor transform, 43
  inverse wavelet transformation, 60–64
  one-dimensional continuous Fourier
    transform, 37–38
  orthogonality, 47
  partial code examples, 49–64
  two-dimensional continuous Fourier
    transform, 38
  wavelet transform, 44–45
  wavelet transformation, 60–64
Image restoration, 37, 87–96, 102
  based on degradation model, 89–90
  constrained conditional
    restoration, 90
  geometric rectification, 93–96
  image degradation model, 87–88
  inverse filtering, 91
  pixel intensity confirmation, 96
  spatial geometric transforms, 94–95
  unconstrained conditional
    restoration, 89
  Wiener filtering, 92, 93
Image segmentation, 84, 135–166
  adaptive method, 138
  band thresholding, 136–137
  border tracing, 148–149
  border-tracing method, 161–166
  edge-based segmentation, 140–151
  edge image thresholding, 140–142
  edge relaxation, 143–147
  exercises, 155–156
  histogram-based thresholding,
    137–138
  Hough transform, 150–151
  image recognition, 155
  iterative thresholding, 138–139
  Mode method, 138
  optimal thresholding, 138–139
  optimal thresholding segmentation,
    158–160
  partial code examples, 158–166
  region-based segmentation, 152–154
  region-growing method, 152
  region-merging method, 153
  region split-and-merge method, 154
  semithresholding, 136–137
  thresholding, 136–139
Image smoothing, 71, 73–79
  Butterworth low-pass filtering, 79
  convolution kernel, 48
  frequency-domain methods, 78–79
  Gaussian filtering, 75
  ideal low-pass filtering, 78
  linear, 73
  median filtering, 76
  methods, 80
  neighbourhood-averaging methods,
    73–74
  nonlinear, 73
  spatial-domain methods, 73–77
  threshold-averaging methods, 75
  trapezoidal low-pass filtering, 79
  weighted median filtering, 77
Image storage formats, 17–28
  BMP format, 17
  GIF format, 19
  JPEG format, 18
  RAW format, 17
Impulse noise, 88
Impulse-response functions, 72
Information theory, 223
Intensity, 8
  function, 14
  grey-scale image, 36, 48
  images, 84
  matrix, 36, 48
International Telegraph and Telephone
    Consultative Committee
    (CCITT), 220

Inverse discrete Fourier transforms, 39
Inverse filtering, 91
  image preprocessing techniques, 91
  image results, 92
Inverse Fourier transform, 37
  one-dimensional cases, 38
Inverse restoration, 92
Inverse wavelet transformation, 60–64
Isotropic diffusion, 100
Isotropic heat conduction model, 98
Iterated functional system (IFS), 217
Iterative algorithm, 211

J

Joint Photographic Experts Group (JPEG), 200, 212, 220
  compression, 201
  DCT-based method, 221
  image storage formats, 18
  predictive method, 221
  standards, 221

L

Lagrange multiplier method, 89, 102
Laplace equation, 97
Laplacian operator, 84, 85–86
Laplacian template, 85
Light exposure, 66
Limiting linear transform, 70
Linear filtering method, 75
Linear image smoothing, 73
Linear shift-invariant (LSI), 35, 72, 88
Linear transformations, 69
  limitation, 70
Local operation, 29
  histogram-based thresholding, 138
Lossless compression, 199, 201–206
  Huffman encoding, 201–205
  runlength encoding, 206
Lossy compression, 199, 207–219
  fractal compression, 217–219
  predictive compression methods, 207
  vector quantisation, 208–211
  wavelet compression, 212–216
Low-pass filtering, 78, 79
LSI. see Linear shift-invariant (LSI)

M

Mathematical morphology, 167–180, 184
  binary erosion, 186–194
  binary images morphology, 169–177
  binary morphological operation, 171–175
  binary-scale images, 167
  binary skeleton, 195–198
  dilation operations, 171–172, 178–181
  erosion operations, 173–174, 182
  exercises, 185
  grey-scale images, 178–198
  grey-scale images applications, 183–184
  hit-or-miss transformation, 175, 176–177
  opening and closing operation, 175
  operations, 178–182
  partial code examples, 186–198
  sets and elements, 168
  sets operations, 168
  sets relationships, 168
  set theory, 168
  skeleton method, 177
  thinning and thickening, 176
Median filtering, 77
  image application, 74
  image preprocessing techniques, 76, 112–117
  spatial domain methods, 73
Mode-histogram-based thresholding, 138
Moving Pictures Experts Group (MPEG), 200, 220
  B-frames, 222
  frame types, 223
  I-frames, 222
  P-frames, 222

N

National Standard Committee (NTSC), 19
Neighbourhood-averaging method
  image application, 74
  image blurring, 75
  image preprocessing techniques, 73–74
  spatial domain methods, 73

Neighbourhood operations, 29
  histogram-based thresholding, 138
Neighbouring pixels position, 82
Noise
  degradation system, 92
  Gaussian, 88, 99–100
  impulse, 88
  ratio, 92, 200
Noisy images, 104
Nonlinear image smoothing, 73, 76
Nonmaximal suppression edge pixels, 141
NTSC. see National Standard
  Committee (NTSC)

O

Object recognition, 84
One-dimensional continuous Fourier
  transform, 37–38
One-dimensional fast Fourier
  transform, 41
One-dimensional Fourier transform, 39
Optic system image differences, 71
Optimal thresholding, 139
  image segmentation, 158–160
Orthogonal function, 47
Orthogonality, 47
Orthogonal property, 47
Orthogonal transformations, 48

P

PAL. see Phase Alternate Line (PAL)
Parabolic equation, 103
Parseval's theorem, 40
Partial code examples, 21–28
  24-bit colour image conversion to red
    channel image, 22–24
  8-bit colour image to grey-scale
    image conversion, 25–28
  8-bit grey-scale image conversion to
    red channel image, 22
  border-tracing method, 161–166
  image compression, 226–242
  image preprocessing techniques,
    108–134
  image processing tools, 49–64
  image segmentation, 158–166
  mathematical morphology, 186–198
  optimal thresholding segmentation,
    158–160

Partial differential equations (PDE),
  97–134
  8-bit grey-scale image histogram,
    108–112
  continuous models, 103
  diffusion-based methods, 97
  diffusion-based models, 98–101
  discrete formats, 103
  exercises, 105–107
  fourth-order, 114–134
  heat conduction model, 98–101
  image-processing model, 97, 103, 104
  median filtering, 112–117
  partial code examples, 108–134
  second-order, 114–134
  Sobel operator gradient image,
    118–123
  total-variation-based models, 97
  TV-based models, 102
PCM. see Pulse code modulation (PCM)
PDE. see Partial differential
  equations (PDE)
Peak-to-peak signal-to noise- ratio
  (PSNR), 200
Period frequency, 1
Perspective distortion, 95
P-frames, 222
Phase, 2
Phase Alternate Line (PAL), 19
Photographing procedures
  brightness, 66
  light exposure, 66
Pincushion distortion, 95
Pixel brightness
  contrast stretching, 69–70
  histogram equalisation, 66–68
  histogram image enhancement, 66–68
Pixel intensity
  confirmation, 96
  rate, 219
Pixel neighbourhoods, 30
Pixel points, 141
Plancherel theorem, 40
P-M model, 99
  diffusion, 101
  restored images, 101
Point operations, 29
Point-spread functions, 72
Predictive encoding system, 222
Prewitt operator, 83, 86

Probability density sketch map, 66
Processing transfer functions, 91
PSNR. see Peak-to-peak signal-to noise- ratio (PSNR)
Pulse code modulation (PCM), 207

R

Range blocks, 217
 image, 218
RAW format, 18
Red, green, and blue (RGB), 9
 channels, 16
 colour model, 9
 to HSI model conversion, 14
 images, 14
 space, 16
 YUV model, 12
Region-based segmentation, 152–154
 growing method, 152
 merging method, 153
 split-and-merge method, 154
Region-growing method, 152, 156
Region split-and-merge method, 154
Representative point, 170
Resolution
 grey-scale images, 6
 images, 6
RGB. see Red, green, and blue (RGB)
RMS. see Root-mean-square (RMS)
Roberts operator, 83, 86
Root-mean-square (RMS), 200
Row-by-row order, 77

S

Salt and pepper noise, 88
Sampling process, 4
Saturation, 8
 function, 14
Scaling coefficients, 216
Scaling functions, 213, 216
Second-order partial differential equations
 image restoration, 114–134
Segmentation methods, 136. see also Image segmentation
 edge-based, 140–152
 region-based segmentation, 152–154
 thresholding, 137
Semithresholding, 136

Set theory, 168
 binary images, 169
Shift-invariant system, 72
Signal combination, 3
Signal-to-noise ratio (SNR), 92
Skeleton method, 177
Smoothing. see also Image smoothing
 discrete Gaussian convolution kernel, 98
 Gaussian, 75, 98
 image processing, 65
 linear filtering method, 75
 linear image, 73
 nonlinear image, 73, 76
SNR. see Signal-to-noise ratio (SNR)
Sobel operator, 83, 86
 gradient image, 118–123
Spatial domain, 73–77
 Fourier transform, 38
 Gaussian filtering, 73, 75
 median filtering, 76
 neighbourhood-averaging methods, 73–74
 shifts, 40
 threshold-averaging methods, 75
 weighted median filtering, 77
Spatial geometric transforms, 94–95
Storage. see also Image storage formats
 analogue signals, 3
 image processing, 199
Streaking, 142
Stretching linear transform, 70
Structuring element, 170
Symmetry transformation formulas, 39

T

Television based models, 102
Thinning operation, 176
Thresholding
 averaging methods, 75
 hysteresis, 142
 image preprocessing techniques, 75
 method, 136
 segmentation, 137
Total-variation-based models, 97
Transfer functions, 72
Trapezoidal filtering
 high-pass, 85, 87
 low-pass, 79
True colour images, 9

24-bit colour image
   conversion to red channel
      image, 22–24
   RGB, 10
Two-dimensional continuous Fourier
      transform, 38
Two-dimensional Fourier transform, 39
Two-dimensional functions, 34
   signal images, 1, 31

U

Unconstrained conditional restoration, 89
Undistorted images, 94
Uniform quantisation process, 5

V

Vector quantisation method, 208
Vertical synchronous (VSYNC), 19
Video, 19
Visible light spectrum, 8
Visual system, 10
VSYNC. *see* Vertical synchronous
      (VSYNC)

W

Wavelet compression, 212–216
   images, 212
Wavelet functions, 214, 216
Wavelet transform, 44–45
   continuous wavelet transform, 44
   discrete wavelet transform, 45
   image processing tools, 60–64
Weighted median filtering, 77
Wiener filtering
   image preprocessing techniques, 92
   image restoration, 93
   random noise, 94

Y

YIQ colour model, 10–11
   components, 11
Y-K diffusion model, 99, 101, 103
YUV model, 11–12
   colour, 11
   RGB model, 12
   transmission, 11
   video encoding, 11